# INNOVATION & IMITATION FOR NATIONS

THE TECHNOLOGICAL GAP SHOCK AND NATIONS' URGE TO IMITATE & COPYCAT

BY
PROF. MOHAMMED AHMAD S. AL-SHAMSI

**Innovation & Imitation for Nations**

The technological gap shock and Nations' urge to imitate & copycat



1st Edition

Published Date 22/2/2022

ISBN 978-1-7346287-9-1

Book editing/translation by Amgad Abouamasha
Book cover design by Mexelina
Book layout by Pankaj Runthala

To My Father,
Oh God, have mercy on my father
and make his grave kindergarten
from the garden of paradise

# Acknowledgment

To all those who supported me in the process of writing this book during the last two years. I am grateful to Prof. Ahmad Mohammed Al-abdulgader for his valuable and unlimited support. I am also grateful for King Abdulaziz City for Science & Technology (KACST).

# TABLE OF CONTENTS

## PART THREE

### Debunking European Innovations From The Fourteenth to The Nineteenth Centuries

## PART FOUR

## USA: The Country That Rose on Imitation and Copycating

# PREFACE

"I have been wondering *why recent innovations came from a specific country or geographic area: the USA and Japan today and Europe before them?*". It took me two years in research to answer this question. I made notes for myself to keep track of selected innovations before deciding to publish a book summarizing my findings. However, I divided the topics into three books to maintain the coherence and sequence of the content. This book is the first of the trilogy.

Since nations benefit from innovations to revive their economies, I have delved into economic theories that have attempted to explain innovation and its role in the growth of nations. However, I found them, though numerous, unable to explain the role of innovations in the economic system, as I mentioned in this book.

Accordingly, I went deeper to search for historical innovations thousands of years ago to find a more precise answer. I reviewed the innovation of Chinese silk and porcelain, Pharaonic kohl,

# PART ONE

# KNOCKOFF VERSUS INNOVATIVE ECONOMY

# CHAPTER ONE

However, innovation is costly as it involves a massive trial and error cycle to have innovative products. Innovators or developers need time and accumulated experiences and an infrastructure geared to development and innovation, which is rarely accessible to small and medium-sized enterprises (SMEs) that do not have the potentials for research and development (R&D), laboratories, and enough profit margin that can be used for R&D while large corporations have an edge in this respect (5).

This applies to countries whose first resort is usually imitation, i.e., emulating competitors' innovation in the markets, be they domestic or international. The strict innovation protection laws increase the burden on these countries and economies, especially the developing nations, and prevent them from infringing on others' intellectual property. Rigid implementation of intellectual property legislations harms the developing countries in favor of the developed industrial countries (6-8). Then, they would have four options, namely 1) to stop imitation and cave to the laws thus incurring an economic loss, 2) to sell such copycat products in the black markets, 3) to embrace innovation to compete, or 4) to wait the end of the legal protection of patented innovation, which constitutes a monopoly up to 20 years in most countries today, and then begin imitation of successful innovations.

Innovation is not always the solution because many countries lack the infrastructure conducive to innovation in all or specific areas. Establishing a national infrastructure that enables innovation, even in one specific field, requires decades and may fail. Accordingly, imitation is undoubtedly easier than innovation, less costly, less time-consuming, and less effort consuming as regards unsuccessful experiments to have a new product (9). However, imitation is not as easy as it seems. The imitator should have some talent to copycat an innovation, even if it exists in front of him. He needs some experiments to reach what others have achieved. This process (as well) requires an "imitation" infrastructure, whether for companies or nations. It is a strategy many companies adopt and succeed in producing a product that resembles an innovative product. The imitation strategy increases production and profits for companies (9, 10). This applies to nations whose self-sufficiency may also cause the knockoff products to replace the imported, innovative product. The United Nations Conference on Trade and Development (UNCATD) considered the imitation as a channel for technology transfer to developing countries (11). While imitation helps imitative countries or companies, it is also beneficial for innovative ones because it is per se an incentive for the innovator to continue innovation and produce other new products to conquer the market anew.

Studies found that 60% of the innovations that entered the market were simulated within four years (12). As an example, the Soviets produced the nuclear bomb four years after the Americans first launched it in 1945. Increasing their economy's ability for technology absorption, the Japanese managed to reduce this period to a year in 1960s-1970s, according to Tilton (13). However, the technical complexities and imitator's readiness determine the average time for one innovation or another. It took three years to reproduce compact discs from one country to another while simulating the phonograph[1] technology required thirty years (9).

In short, no matter how high the country's capacity for technology absorption or imitation, the imitation process of new products requires time. Some sources indicate that the average time for technology absorption is four years (i.e., the period before competitors learn the mechanism of imitating the successful innovation) during which the innovative country or company can get the head start in entering the domestic or foreign markets, establish itself there, and reaping profits before the copycat products and competition enter the same markets even in the absence of the laws and mechanisms for protecting intellectual property. In their book, *The Knockoff Economy: How Imitation Sparks Innovation,* Raustiala and Sprigman demonstrated the benefit of imitation on innovative companies

1 A phonograph is an instrument for recording and reproducing sound.

and countries by studying markets not covered by intellectual property regulations, thus not preventing the emergence of the counterfeit products following the success of the innovative example or model in the markets, including clothing, fashion, arts, cinematographic, and food. Although the regulations do not protect innovation in these markets, they have been thriving and creating new products that meet the consumers' diverse needs. Contrary to the claim, imitation did not end innovation, but was rather a catalyst for its continuation (14).

Back to the benefits of imitation for the knockoff company, country, or factory, apart from the material profits accruing from creating copycat products and services, imitating countries may over time become innovative, as imitation may be a step towards innovation (15). This was the case of Japan after more than 20 years of imitation starting 1960 and South Korea in the eighties and nineties (16, 17). This has been happening with China as of 2000 (18). The imitation strategy may be beneficial and temporary, after which the imitating nation turns into an innovative one. Imitation may be the first stage of national industrial and technical progress. This may apply to companies. Microsoft has kept pace with Apple and increased its economic returns even though Apple had produced and marketed its Macintosh operating system first. Apple filed a complaint with Microsoft in 1988 alleging the theft and reproduction of 189 components of the Macintosh operating system (19). The US car companies that succeeded Ford obtained higher market shares though Ford had introduced cars first to the US market.

Just as innovation is beneficial to the company, country, and consumer, imitation also benefits the innovator (in terms of creating competition). Nuruzzaman indicated that exporting innovative products to markets in new countries other than those that innovated them was an additional incentive rather than a discouraging factor for innovation despite fears of imitation or flawed intellectual property protection in such countries (20).

Tawfik Dalgic has studied the positive impact of imitation in fifty significant companies in Turkey and proved that the imitation strategy of Turkish companies led to economic success and growth in companies' market shares from the spare parts industry to electronics and from building materials to chemicals (21).

Every innovation that succeeds on the market inspires countless imitators to emulate it and a desire to gain market share. Theodore Levitt states, "In fact, imitation is endemic. Innovation is scarce" (22). He also coined the term Reverse R&D, which expresses the R&D effort to devise a product that has already been invented and marketed. Reverse R&D aims to identify ways to reach the existing innovation. The private sector invests a lot in the reverse R&D process, whose goal is sometimes not profit but only survival in the market. Therefore, companies' imitative ability sometimes becomes necessary to ensure survival. The companies that cannot innovate or imitate are doomed to extinction. This also applies to nations. Therefore, reverse R&D is an active and common strategy even in the most advanced countries and in the most profitable and capitalized companies. Intellectual property rights do not protect all kinds of innovation. Some successful companies, and nations, divide R&D into two parts: innovation and imitation. The former monitors all competitors' products on the market or even before entering the market and try to imitate them quickly to gain a foothold versus the innovative competitor. It is an effective research strategy to ensure that opponents do not outpace and reduce the risk of losing market shares in their favor. It would also be prudent for the knockoff company not to move until after the new innovative and competing product proves successful on the market because most innovative products do not see the light, and those that do not enter the markets and have demand from consumers fail and fade away. Thus, imitation, or reverse R&D should kick off before the innovative product succeeds and attracts consumer demand. This also applies to countries, even in armament technologies. The simulation process should begin after other countries' weapons prove successful. By reviewing the life cycle of any innovative product on the markets, the timing of imitation must be in the first stage before it reaches the peak consumption or the rapid ascent stage. This requires the imitator to exert an effort to intensively monitor competitors and markets, track all new products, and monitor their progress. Accordingly, a mechanism or strategy must be adopted to determine the appropriate timing for kicking off the reverse R&D process (23).

Imitation is always related to successful innovations; wherever there is a successful innovation, there are those who seek to imitate it in a symbiotic process beneficial to both parties even in the total absence of intellectual property protection systems. However, we do not call for the absence of intellectual property rights. Intellectual property protection laws and mechanisms provide an additional incentive to the innovator, who incurs a lot in trial and error until he reaches the desired innovation. This incentive allows the innovator, be it an individual or a company, reap profits for a specified period that does not exceed 20 years during which the innovator maintains the solo right to prevent any other entity or individual from unlicensed imitation of his/its innovation. The time-limited monopoly temporarily prevents imitators except through black markets, which unfortunately the system does not cover. These gray markets are available and widespread. Intellectual property systems that prevent imitators also stimulate the imitators to go beyond the imitation stage and think of something developed with better specifications and characteristics than those of the innovation owned by the competitor. Therefore, intellectual property systems are a catalyst for innovation for both parties. The imitator has no place except by being innovative, which encourages competition and makes the innovation market thrive.

However, creating an innovation infrastructure is as hard as creating a knockoff infrastructure, but one is more complicated. This, in turn, raises the costs of innovation and makes it unaffordable and unavailable to the largest segment of companies, thus keeping them outside competition. This is an economic problem that needs a solution within the same country. When comparing large companies with SMEs, large companies are often the innovators that encounter no or minimal competition from SMEs (9). This concept also applies to major industrial countries compared to developing countries that lack high technical capacity, advanced R&D institutions, qualified human capital, and enough funding to support R&D projects. This makes developing countries (or it is in their interest even if they do not do so) turn a blind eye or be easy in applying intellectual property protection systems and reducing the control of their implementation procedures (24). Even if legitimized, intellectual property systems are not applied effectively thus encouraging counterfeit product markets, which are flourishing with consumers. Later in this part, we will list some of these (in) famous markets.

In 2017, the "Counterfeiting Report" website published that China, the second-largest economy in the world, produced 80% of the global counterfeit products[2] (25). Forbes 2018 forecasts indicated that the global market for counterfeit products would reach US$2.8 trillion by 2022 and create more than 5,000,000 jobs (26). The report "Trends in Trade in Counterfeit and Pirated Goods" published by the Organization for Economic Cooperation and Development (OECD) in 2019 indicated that the value of the counterfeit goods market exceeded half a trillion dollars in 2016, representing 3.3% of the world trade market (27).

The role of local companies in developing countries is limited to imitation as foreign companies own innovations, even if intellectual property systems protect them, due to the inability to innovate and resorting to black markets to manufacture their products and gain some profits. The entry of the foreign product would be positive for local companies in developing countries (after the end of the innovation regulatory protection period). It introduces local companies to modern technology and motivates them to emulate, copy, or imitate these products,

2 It includes the People's Republic of China and Hong Kong as part of Greater China.

especially those that do not contain advanced complex technology whose reproduction infrastructure would need decades to establish (28). The strict systems that protect intellectual property in developing countries reduce imitation and lead to innovation because imitation here may be a first step or a stage towards establishing an imitation infrastructure to reach the innovation stage one day.

Tightening intellectual property laws in developing countries that import their products is a grave economic mistake. However, they must strictly protect local innovations from domestic or foreign competition. The process of enacting laws to protect intellectual property involves discrepancies of the legislator must be aware not to follow the advanced industrial countries' whims, restrict local companies at home, and do not remain lax in a way that wastes the innovation right of local companies. The legislative process of intellectual property laws requires a careful balancing to suit each country according to its capacity, technical readiness, and institutional capabilities for R&D and innovation. The legislator must stand with and defend the local companies, as they are the nation's wealth, even if they are private in which the State does not own a share. The company (even if it is private) is a building unit in the country's economic system. Accordingly, ignorance of the legislator in developing countries about how, when, and in what case intellectual property laws are used will cause the developing countries to miss many economic opportunities and kill their future ambitions.

# References


1. J. Fagerberg, M. Srholec, B. Verspagen, The Role of Innovation in Development. *Review of Economics and Institutions* **1**, (2010).
2. A. Pyka, Prettner, K., Economic Growth, Development, and Innovation : The Transformation Towards a Knowledge-Based Bioeconomy. *Bioeconomy*, (2018).
3. A. M. Pece, O. E. O. Simona, F. Salisteanu, Innovation and Economic Growth: An Empirical Analysis for CEE Countries. *Procedia Economics and Finance* **26**, 461-467 (2015).
4. L. Diaconu, paper presented at the ERSA conference papers ersa, 2011.
5. A. Pierre, A.-S. Fernandez, Going Deeper into SMEs ' Innovation Capacity: An Empirical Exploration of Innovation Capacity Factors. *Journal of Innovation Economics & Management* **25**, 139-181 (2018).
6. A. Sattar, T. Mahmood, INTELLECTUAL PROPERTY RIGHTS AND ECONOMIC GROWTH: EVIDENCES FROM HIGH, MIDDLE AND LOW INCOME COUNTRIES. *Pakistan Economic and Social Review* **49**, 163-186 (2011).
7. N. N. Seeratan, Negative Impact of Intellectual Property Patent Rights on Developing Countries: An Examination of the Indian Pharmaceutical Industry. *Scholar*, 339-377 (2000).
8. T. Kazutaka, Negative Effects of Intellectual Property Protection: The unusual suspects? *Research Institute of Economy, Trade and Industry (RIETI)* **Discussion papers** (2012).
9. O. Shenkar, Copycats: how smart companies use imitation to gain a strategic edge. *Strategic Direction* **26**, 3-5 (2010).
10. G. Schewe, Imitation as a strategic option for external acquisition of technology. *Journal of Engineering and Technology Management* **13**, 55-82 (1996).

11. UNCTAD, "Transfer of Technology and knowledge sharing for development Science, technology and innovation issues for developing countries," *UNCTAD CURRENT STUDIES ON SCIENCE, TECHNOLOGY AND INNOVATION* (United Nations Publication, 2014).

12. E. Mansfield, M. Schwartz, S. Wagner, Imitation Costs and Patents: An Empirical Study. *Economic Journal* 91, 907-918 (1981).

13. J. E. Tilton, B. Institution, *International Diffusion of Technology: The Case of Semiconductors.* (Brookings Institution, 1971).

14. K. Raustiala, C. Sprigman, C. J. Sprigman, *The Knockoff Economy: How Imitation Sparks Innovation.* (OUP USA, 2012).

15. P. Aghion, C. Harris, P. Howitt, J. Vickers, Competition, Imitation and Growth with Step-by-Step Innovation. *The Review of Economic Studies* **68**, 467-492 (2001).

16. M. K. Bolton, Imitation versus innovation: Lessons to be learned from the Japanese. *Organizational Dynamics* **21**, 30-45 (1993).

17. A. E. Fakir, in *Management of Technology Innovation and Value Creation.* pp. 275-292.

18. K. Zhou, Innovation, Imitation, and New Product Performance: The Case of China. *Industrial Marketing Management* **35**, 394-402 (2005).

19. L. Dormehl, 'Today in Apple history: Microsoft gets sued for ripping off Mac OS. 2020.

20. N. Nuruzzaman, D. Singh, C. Pattnaik, Competing to be innovative: Foreign competition and imitative innovation of emerging economy firms. *International Business Review* **28**, 101490 (2019).

21. T. Dalgic, Imitative Innovations: A Product Strategy From a Newly Industrialised Country - The Turkish Case. 92-95 (2015).

22. T. Levitt, Innovative Imitation. *Harvard Business Review*, (1966).

23. C. Karlsson, *Innovation adoption and the product life cycle.* Umea Economic Studies (University of Umea, 1988), vol. 185.

24. E. Helpman, Innovation, Imitation, and Intellectual Property Rights. *Econometrica* **61**, 1247-1280 (1993).

25. Counterfeiting Report (2017).

26. W. Shepard, in *Forbes*. (Forbes, 2018).

27. OECD/EUIPO, "Trends in Trade in Counterfeit and Pirated Goods," (OECD & EUIPO, Paris, 2019).

28. R. E. Caves, Multinational Firms, Competition, and Productivity in Host-Country Markets. *Economica* **41**, 176-193 (1974).

# CHAPTER TWO

## CHAPTER TWO

# IMITATION OF TRADEMARK OF INNOVATIVE PRODUCTS AND THE BOOM OF BLACK MARKETS

The call for inaction in implementing intellectual property systems in developing countries is not the same as allowing infringement of trademark ™, which is a radically different issue. The trademark™ of the innovative product must be protected in all ways to ensure that the consumer is not misled and deceived into buying counterfeit or fake products instead of the original trademark™ products (1). Complacency in this matter helps and encourages fraud and would trigger internal economic corruption. It would also deter foreign companies from entering the local markets in developing countries in which they would lose their products' rights and cannot protect them.

Here, we will differentiate between imitating an innovation to manufacture a knockoff product with a new trademark ™ and producing a copycat product with the same trademark™ of the innovative company owning this trademark™. In the former case, the imitated product copies (or mimics) the same technical skill and innovative know-how of the innovative product to be imitated but with a new trademark™, which may one day outpace or compete with the original, innovative product. The latter case is an outright fraud and deceit to the consumer; it harms the innovator who has the right to the trademark™ (2). We warn against complacency in this regard because of its long-term adverse effects on nations' economies and domestic markets.

When Japan started in the 1960's and Korea in the 1980's, they knew they were (in)famous for imitation, copying, and knockingoff the innovations of some industrially advanced countries (3). However, Japan and Korea were designing new trademarks™ and placing them on the counterfeit products so that the consumer could recognize their products. The cost difference incentivized the consumer who knew the product was of lower quality compared to the original innovative one (4). Moreover, the legal protection period of some innovations expired, and the imitator had the legal right to produce them. As for the trademark ™, it would remain for the first product and does not expire even after the termination of the legal protection period except in three cases: liquidation of the company or discontinued payment the trademark™ fees in the country of origin or not registering it in the developing countries to which the product may reach indirectly through other intermediate markets. In general, the protection of trademark™ rights takes precedence over the protection of innovations, though both are important because the monopoly of the trademark is harmless even in the long run, while the long monopoly of industrial innovation harms the innovator and the consumer. It discourages the innovator's ambition to produce other innovations if his innovation does not have competitors in the markets while the consumer will incur the product's higher costs.

To encourage imitation for foreign innovation among local companies in a vital sector, developing countries may impede the registration or grant of intellectual property rights to foreign companies in their territories. In this way, the activity of local knockoff companies shall not violate domestic law. Although exporting these counterfeit products to other countries, especially those in which the innovation process has taken place, will be difficult, developing countries may benefit from taking a step forward to establish and finalize the infrastructure for this knockoff industry. Once the innovation legal protection period in other countries (including the country of origin of the innovation) ends, these developing countries can legally market their knockoff product in those countries. The first period, which allows imitation in the local markets, would be an internal incubation period, followed by the launch period, which begins with the end of the innovation legal protection period in the country of origin and other countries in which the innovation is patented.

This knockoff product may invade international markets and generate a sizeable profitable return, especially if developing countries benefit from low-wage labor or incentive tax exemptions/credit in some vital sectors. In this case, the consumer in the developed countries that invented this technology would also benefit when he obtains the product either at a lower price or even by providing multiple options when purchasing.

Accordingly, legislators in developing countries must draw up a list of vital innovations in the sectors in which a robust national industry is envisaged and then prohibit, obstruct, or restrict granting intellectual property in these products to foreign companies. This awareness among legislators in developing countries will positively impact the national economy and not violate international law. These countries have the right to cater for, defend, and revive their economies through laws and legislation, provided that the process is not arbitrary in such a way that would prohibit or obstruct intellectual property for all foreign companies. Instead, it must be codified and regulated as per a list of national priorities and in specific sectors in which the developing country would compete one day. This also applies to the advanced industrial countries when competing in specific technical fields. They must protect their respective national economies with such procedures and legislation.

Should China rigidly prohibit all kinds of intellectual property infringements down to a hypothetical zero percentage of sales, the Chinese economy would indeed decline, given that the report mentioned above states that 80% of the counterfeit products in the world are made in China. However, this does not deny that the Chinese economy is the second-largest globally (5). Accordingly, we must differentiate between positive imitation and legitimate copying and negative and illegitimate one, both of which coexist in China.

Not putting a new or different trademark ™ from the original, innovative product at the time of imitation would severely damage the imitative company or factory in the long run. It is a major cause for not achieving any future success because the copy is hidden from the consumer and does not carry any trademark™ that distinguishes it. However, if the imitative company has a distinctive trademark™, there would be future demand for this trademark™ for several reasons, such as the low price or the easy access of a product to local markets thanks to distribution networks or advanced logistical mechanisms or other reasons beyond the control of the imitative company, such as the collapse of the innovative company in its country due to legal cases, misfortune, or inter-State dispute that prevents the original, innovative product from reaching other countries.

In all these cases, a counterfeit product with its trademark ™ may invade a market, even if it was of lower quality in the eyes of consumers initially, or if consumer confidence is still lacking even if the quality is high. A knockoff product with its own trademark™ takes time to gain consumer confidence. However, these multiple internal and external conditions will make the knockoff product with its own trademark™ to assume a position on the market. In the case of imitation of a trademark™ product, the imitator will not have a chance for success in the future, let alone it is outright fraud. This imitative company will not turn to the international markets if it succeeds in the local markets except through the illegitimate unregulated black markets.

Hence, companies that try to emulate, copy, or imitate innovative products with proven success in the market must enter with their own trademark ™ and be satisfied at the outset with their limited position in the market and low market share until it is time and opportunities arise to increase the market share or develop the knockoff product over time and produce a commodity whose specifications and properties exceed the original innovative product, which will grant it a competitive advantage and raise its position in the market. The bottom line is that the damage of imitating the trademark™ in the long run is a disadvantage of the original innovative product whose name will remain prominent in the market because the imitator would be spreading the innovator 's trademark™. In this case, the imitator promotes the original innovative company's trademark™ due to lack of future awareness. For example, the Boeing 747 commercial aircraft was marketed in 1968, that is, more than 50 years ago. This product, which involved a set of innovations when launched, no longer has intellectual property rights. Today, any company can simulate this technology and re-launch it in the market, whether locally in specific developing countries or even export it to the manufacturing countries, without any legal disputes or infringement on intellectual property provided that the Boeing logo is not placed on the new knockoff aircraft. It is noteworthy that the fleets of major international commercial airlines to this day still contain the Boing 747. The total number of that aircraft still in use is estimated at 438 units. Boeing sold 1,558 aircraft until 2018 (6). It is also worth noting that the Boeing 747 is an improved model of the 707 introduced in 1958.

The development of its prototype took a team of 4,500 engineers led by Joe Sutter twenty-nine months and cost one billion dollars at the time. When launched, it was the largest civil aircraft of 231 feet long and a capacity of 360 passengers (7,8). Boeing is still very active in innovation, with more than 1,400 patents granted last year by the United States Patent and Trademark Office (USPTO) (9). The company is trying to maintain its market position by making minor modifications to the original model and protecting it with a new patent. Over the past 60 years, the company offered several improved Boeing 747 models, from the 747-100 to the 747-8F (7, 8).

To avoid any doubt, innovations must be disseminated and benefit humankind rather than monopolized. The protection granted by countries through patents is just an incentive in the form of a limited-time monopoly to reward the innovator by allowing him to reap some profits from his idea/ product for a temporary period, after which ownership of intellectual production becomes available for others. The end of the legal monopoly period allows competing companies to emulate and imitate this innovation, a legal right guaranteed by regulations in developing or developed countries. Any society and its legislators must not lose sight of the importance of making innovations beneficial to all. If there were an eternal (permanent) monopoly on innovation, this will discourage the innovator, as he may be satisfied with what he has achieved. This will be also more discouraging for competitors by depriving them in the future of the areas of success achieved by a competitor. In the case of developing countries, it would more discouraging if they are denied all progress on the grounds that all such innovations are patented for companies in foreign countries in developed nations and cannot be reproduced forever.

In her book *Faked in China : Nation Branding, Counterfeit Culture, and Globalization*, Fan Yang noted that the market for counterfeit products reached US$250 billion in 2007. Many vibrant and frequented markets are dedicated to counterfeit products in today's world (10). Below, we list some of the most (in)famous of these markets.

**Table (1): List of the most (in)famous counterfeit goods markets (wholly or partially) in the world.**[3]

| Name Market | City | Country |
|---|---|---|
| **Canal Street** | New York | USA |
| **K Street Corridor** | Washington | USA |
| **Silk Street Market** | Beijing | China |
| **Petaling St.** | Kuala Lumpur | Malaysia |
| **Night Bazar** | Pattaya | Thailand |
| **Mike Shopping Mall** | Pattaya | Thailand |
| **MBK Shopping Mall** | Bangkok | Thailand |
| **Kusa Dasi Bazar** | Kusa Dasi | Turkey |
| **Shanghai S&T Metro 2** | Shanghai | China |
| **AP Plaza** | Shanghai | China |
| **Carment Wholesale center** | Guangzhou | China |
| **Luohu Commercail center** | Shenzhen | China |
| **Saigon Square** | Ho Chi Minh | Vietnam |
| **Ben Thanh Market** | Ho Chi Minh | Vietnam |
| **Mongkok Ladies Market** | Hong Kong | Hong Kong |
| **Karama Market** | Dubai | UAE |

3 As per the information available to the author, noting that official sources do not endorse the availability or existence of such markets, nor allow the sale of counterfeit products. Its existence is real. Its approval is another legal matter.

Because these are counterfeit and copycat product markets, reports do not accurately rank them in size or activity, rendering available information somewhat speculative rather than a reliable source to be cited in a book. Although regulations in some countries prohibit that, the strength of consumers' demand for counterfeit products is very high, which encourages the creation and survival of such markets (4). As for the consumer, he goes to these markets not in search of the original product but rather in pursuit of the counterfeit product while fully aware of that fact. Here, the seller does not deceive or mislead the consumer but rather fulfills his desire for high-value goods at a meager price (11). The victim here is the producing trademark company or the company that holds a valid innovation right (under legal protection) and not the consumer (2). The Organization for Economic Co-operation and Development (OECD) indicated that about 59% of consumers in the markets know that the product is counterfeit before purchasing. Yet, they opt for the price difference, willingly waiving part of the quality in the process (12).

Some press and field reports on counterfeit products referred to well-known branded goods at less than 1% of their original value. It is also difficult to differentiate between the shape of the original and the knockoff product, especially clothing, luxuries, and shoes. When asked that the Alibaba platform is (in) famous for counterfeit products, which is an infringement of others' intellectual property, Jack Ma, founder of the Alibaba Group, which specializes in providing electronic platforms for the sale and delivery of goods to consumers, replied in a TV interview that it is challenging to know whether the product is a counterfeit or not, due to the high accuracy of imitation of some producers, adding that the tests to determine the difference did not often succeed. This fact has already been proven by academic studies on the subject (13-15).

A report of the OECD and the European Union Intellectual Property Office (EUIPO) in 2019 entitled *Trends in Trade in Counterfeit and Pirated Goods* indicated that China is at the fore in the production of counterfeit goods and covers more than 50% of the relevant global market, followed by Hong Kong (27%), Turkey, Singapore, Germany, India, and Thailand. The most affected companies were from the USA (24%), France (17%), Italy (15%), Switzerland (12%), Germany (9%), and Japan (7%) (12).

Hong Kong, Singapore, and the UAE are the most transit stations for transporting counterfeit goods from the producing countries to the final markets for sale. Most counterfeit goods were shoes, clothes, leather goods, and watches, with a market share ranging from 10-23%, followed by fake perfumes, jewelry, and toys, which range from 3-6% of the market (12).

These trademarks, which are infringed upon in knockoff markets, are a fortune in global trade. The value of global intangible assets [which include trademarks and other intangible assets, such as patents, databases, etc.] was estimated at US$476 billion in 2016 compared to US$198 billion in 2001. These intangible assets represent 70% of the world economy (16).

The value of the Amazon trademark amounted to US$100.7 billion in 2018 compared to US$4.5 billion in 2000. Accordingly, it has grown more than twenty times during this period. The value of the Apple trademark amounted to US$214.5 billion in 2018, up from US$6.6 billion in 2000 (17).

These figures show the rapid growth in the trademarks' value ; the concept of the trademarks' high value in developing countries has exceeded tens of billions. As per the Forbes 2020 World's Most Valuable Brands report, the value of the Korean company Samsung trademark exceeded US$50 billion, the Japanese company Toyota 's US$41 billion, and Honda 's US$24 billion (18).

The Brand Finance China 500 report indicated that the brand value of the top 100 companies in China is worth US$1.486 trillion and has doubled more than eight times in just a decade, up from US$175 billion in 2010 (19). The value of the Chinese company Huawei has reached US$62.3 billion

(20). The Brand ZTOP report showed that the Alibaba brand value reached US$152.2 billion in 2020 (21).

However, the brand value information for any company involves severe discrepancy due to the different mechanisms and methods for evaluation of intangible assets or the inclination of one of the evaluating parties or entities in favor of a specific brand or country. However, what matters is that what is being imitated equals huge fortunes, some of which equal countries' budgets for decades. The ability of some developing countries with limited resources to manufacture a brand is difficult to realize in a short period, and it takes decades to build a brand name that derives its reputation from consumer satisfaction with the product quality.

In their book *Global Culture Industry: The Mediation of Things,* published in 2007, Lash & Lury explained that some products had become strong cross-border cultural symbols that affect peoples' cultures. The way to produce these symbols is to create a new culture for the world (22).

While people speak several languages, they pronounce the brand's name in the same way (with a difference in dialects and articulation points). From Latin America to East Asia and from Alaska to South Africa, they pronounce Coca-Cola, Rolex, and Mercedes in the same way. It is then a new culture adopted by humans through consumption and has penetrated the State borders to become part of contemporary languages (though most of them have no meaning in most languages). The brand's strength derives from the symbol's connotations that transcend the language barrier among the peoples. These symbols also carry a memory that embodies consumers' experiences of a good or service (23-26).

Then, it is no wonder that there is a strong desire to imitate (or even attempt to counterfeit) a successful brand product. However, the author advocates imitation, copying, and simulation with limitations that we explain in different parts of the book as a strategy for innovation; he refuses counterfeiting, fraud, and deception. In the former case, copying the product without the trademark may be legitimate, especially if the intellectual property protection period for the innovation has expired. The latter is misleading and deceptive in which the product and its trademark are copied either to mislead the customer, the competitor, or the regulatory authorities. Even if the customer is satisfied, the trademark owner still has a right. Even if the legal rights of the trademark expire, it remains an obituary of the future of the imitated entity. The trademark counterfeiter deprives himself of the future opportunity to acquire market shares for his knockoff product, acquiring a name in the markets that is appealing to the consumer.

## References


1. R. Wilke, J. L. Zaichkowsky, in *Business Horizons.* (1999), vol. 42, chap. 9, pp. 9.
2. A. Mallik, *Ethical Perspective of Brand Imitation on Consumer Buying.* (2020).
3. M. K. Bolton, Imitation versus innovation: Lessons to be learned from the Japanese. *Organizational Dynamics* **21**, 30-45 (1993).
4. H. Mujahed, A Study to Show the Acceptance of Brand Imitation According to the Opinions of Multimedia University International Students-In Malaysia. *The International Journal of Business and Management Research* **VII**, 58 (2014).
5. X. Zhu, Understanding China 's Growth: Past, Present, and Future. *Journal of Economic Perspectives* **26**, 103-124 (2012).
6. CAPA. (CAPA, 2018), vol. 2/2/2021.
7. M. W. Bowman, *Boeing 747: A History: Delivering the Dream.* (Pen & Sword Books, 2014).
8. G. Norris, M. Wagner, *Boeing 747: Design and Development Since 1969.* (Motorbooks International, 1997).
9. T. Alsop. (Statista, 2021).
10. F. Yang, *Faked in China : Nation Branding, Counterfeit Culture, and Globalization.* (Indiana University Press, 2015).
11. C. Viot, A. Le Roux, F. Krémer, Attitude towards the purchase of counterfeits: Antecedents and effect on intention to purchase. *Recherche et Applications en Marketing (English Edition)* **29**, 3-31 (2014).
12. OECD/EUIPO, "Trends in Trade in Counterfeit and Pirated Goods," (OECD & EUIPO, Paris, 2019).

13. E. Penz, B. Stoettinger, Original Brands and Counterfeit Brands - Do They Have Anything in Common? *Journal of Consumer Behaviour* **7**, 146-163 (2008).

14. B. Stöttinger, Penz, E., in *The Customer is NOT Always Right? Marketing Orientationsin a Dynamic Business World. Developments in Marketing Science: Proceedings of the Academy of Marketing Science.*, C. L. Campbell, Ed. (Springer, 2017).

15. E. Triandewi, F. Tjiptono, Consumer Intention to Buy Original Brands versus Counterfeits. *International Journal of Marketing Studies* **5**, (2013).

16. N. Z. Ökten, E. Y. Okan, Ü. Arslan, M. Ö. Güngör, The effect of brand value on economic growth: A multinational analysis. *European Research on Management and Business Economics* **25**, 1-7 (2019).

17. G. Dorfleitner, F. Rößle, K. Lesser, The financial performance of the most valuable brands: A global empirical investigation. *Heliyon* **5**, e01433 (2019).

18. M. Swant, "The World's Most Valuable Brands," (Forbes, 2020).

19. Brand Finance, "The annual report on the most valuable and strongest Chinese brands," *BRAND FINANCE CHINA 500 2020* (Brand Finance, 2020).

20. World News. (World News, 2019).

21. Kantar, "Brandz Top Most Valuable Golabal 75 Retail Brands 2020,» (Brandz, 2020).

22. S. Lash, C. Lury, *Global Culture Industry: The Mediation of Things.* (Wiley, 2007).

23. C. Lury, *Brands: The Logos of the Global Economy* (Routledge, 2004).

24. L. Moor, *The Rise of Brands.* (Berg Publishers, 2007).

25. D. B. Holt *et al.*, *How Brands Become Icons: The Principles of Cultural Branding.* (Harvard Business School Press, 2004).

26. A. Arvidsson, *Brands: Meaning and Value in Media Culture.* (Routledge, 2006).

# CHAPTER THREE

CHAPTER THREE

# FROM RICARDO TO KRUGMAN: FAILURE OF ECONOMIC THEORIES TO UNDERSTAND INNOVATION

In *The Basics of Tax and Economic Policy,* the British economist David Ricardo introduced the concept of comparative advantage in 1817. Ricardo pointed out that a nation's ability to produce goods at a lower price is an advantage compared to other countries producing the same goods. This would make money flow to the former nation in trade, resulting in an accumulated wealth at home (1).

Ricardo compared the advanced industrial countries. Then, the economists realized that the non-industrial developing countries were trying to produce the same goods and products. Compared to the high labor income of the industrialized countries, even if the practical skill and the amount of hard work and diligence are equal, the worker in the industrialized countries receives a higher wage than his counterpart in the developing countries. There remains only the factor of transferring knowledge, skill, or way of the mechanical machine operation to developing countries. When a developing country obtains the technology, it can reproduce or imitate the same commodity at a lower or same quality, adding the advantage of cheap labor cost. Thus, it can produce the same goods at a lower price. As per Ricardo's description, it would have a comparative advantage, which will allow it to adjust the inter-State trade balance.

After analyzing this matter, the economist Raymond Vernon came to develop the Product Life Cycle theory in his work *International Investment and International Trade in the Product Cycle,* published in 1966. Vernon indicated that the high-tech product innovated in a developed country would be exported to developing countries after success in the advanced industrial country. When this product is imported into the developing country and gains a noticeable and stable growing consumer market, the developing country would imitate it as per its capabilities. Attempts would continue until that developing country learns the technical knowledge to manufacture this product to the satisfaction of the national consumer. This product would then replace the imported product.

In fact, over time, and with the development of the local industry in the developing country, it may re-export its products to other countries (both developed and developing), including the manufacturing country of origin of this innovative product (2). Defining the product life cycle was Vernon's contribution to the analysis of international trade.

In 1933, Heckscher and Ohlin developed the Heckscher-Ohlin mathematical general equilibrium model of international trade (H-O model). It built on Ricardo's theory of comparative advantage by predicting patterns of production and international commerce pathways. The model essentially says that countries export the products which use their abundant and cheap factors of production and import the products which use the countries' scarce factors. Export is thus an advanced stage after the success of production at the national level and an effect of production abundance (3). However, the proposed Heckscher Ohlin mathematical model did not capture the production potential of developing countries. It seems that it believed in developing countries' inability to produce advanced technical quality and innovations. Moreover, the model could not accommodate the possibility of transfer of capital and labor between developing and developed countries and vice versa.

The US economist Paul Krugman, the winner of the Nobel Prize in Economics in 2008, developed a theory of international trade that explains the above based on two main factors: technology transfer and innovation. If the rate of technology transfer from developed countries to developing countries increases, the latter's income would increase. When the progress in innovation is stable (i.e., no innovations entering the market) in the developed countries, only technology transfer to developing countries would continue. The per capita income between the developed and developing countries would then reach the break-even point due to the equal opportunity in production for the advanced industries (4).

However, if the innovation of the new products fed into the markets and exported to the developing countries continues in the developed countries, the per capita income would remain high (in developed countries) even with the continuation of technology transfer. The once new technologies would become old and available for production in developing countries. The high rate of innovation in developed countries increases the income gap between developed and developing countries and vice versa for the increasing rate of technology transfer. It is worth noting that Krugman advised developed countries to reduce or hinder the pace of technology transfer from developed to developing countries to maintain a high income, especially in the event of a low rate of innovation for new products in developed countries. I believe that awarding the Nobel Prize to Krugman was due to his explaining how to keep developing countries always of low income by delaying or impeding the transfer of technology to them.

Krugman, Ohlin, and the rest of the economists overlooked that developing countries could one day innovate independently. However, over the past two hundred years, the economic models did not absorb that. Today's most developed industries are produced by countries that economists used to call developing (and some still do) and underdeveloped, such as China, Japan, India, Asian tiger countries, Brazil, Mexico, etc.

For example, the trade balance between Mexico and the USA has recently tilted in favor of Mexico after signing the North American Free Trade Agreement (NAFTA). Likewise, the trade balance between China and the USA tends to favor China.

An analytical view of the trade balance between developed and developing countries used as a criterion for judging the economic maturity and technical development of countries in terms of preference in trade balance and the quality of what is exchanged in terms of technical level shows that, in general, the trade balance began to tilt towards the developing countries during 1980-2000, as we will explain below. The trade balance between the USA and China started to lean towards China in 1993. Exports from China to the USA were estimated at US$17 billion, and imports from the USA to China were estimated at US$10.7 billion, i.e., a trade balance difference of more than US$6 billion in favor of China (5). Exports grew gradually between China and the USA in favor of China up to US$471.8 billion in 2019, and imports to China from the USA are estimated at US$163 billion, i.e., the difference in the trade balance between the two countries is US$308.8 billion in favor of China (6).

The developing countries were accused or described as producing labor-intensive and low-quality goods, according to the description of the economists referred to at the beginning of this chapter. But today, we see that exports from China to the USA include electronics and machines at US$217 billion. These high-tech products imported into the USA from China exceed all US exports to China, which did not exceed US$163 billion in 2019 (5).

If China is an exceptional case, let us look at countries described as developing as per the UN international agreements, including Mexico, India, South Korea, and Indonesia.

The Mexico-USA trade balance in 1995 amounted to US$15.8 billion in favor of Mexico and has been growing up to US$101.4 billion in 2019, with US$358 billion in imports from Mexico compared to US$256.6 billion in exports to Mexico (7).

Since 1985, India has managed to tilt the trade balance with the USA in its favor by US$650 million up to US$23.4 billion in 2019 with US$57.7 billion in exports and US$34.3 billion in imports. Indonesia has followed the same approach since 1985, with a trade balance in its favor amounting to US$3.7 billion. As for South Korea, it has managed, since 1998, to make the trade balance in its favor by US$7.46 billion (US$23.9 billion in exports and US$16.5 billion in imports). This amount increased in twenty years up to US$20.9 billion in 2019 in favor of South Korea (US$77.5 billion in exports compared to US$56.5 billion in imports). (7).

The analysis of the trade balance between industrialized countries and countries classified as developing and non-industrial under international agreements (that are still valid to date), and the theories of economists show that the trade balance began to tend gradually three decades ago in favor of developing countries with a noticeable and steady increase. When we look closely at the analysis of the types of exports to developed countries, we find (as in the example above of China and the USA) they include electronics, machines, cars, and many other advanced industries that contain an innovative dose that distinguishes them from the products of countries classified as advanced (under international agreements).

Accordingly, current economic models do not help us understand international trade in depth nor enable us to understand the impact of innovation on developing countries and the balance of international trade. We need a new economic model that can accommodate these variables in today's world. The ability of countries to innovate cancels out the concept of developing and developed countries. However, international agreements are full of such phrases derived from economists' theories by countries that consider themselves advanced for a period. Nowadays, there is no place for such biased and racist terms. The ability to innovate needs to build an integrated national system capable of innovation. Some countries need to know the mechanisms of simulation and imitation as a way conducive to innovation. Although simulation and imitation are not always necessary to reach innovation, it is an auxiliary or supportive stage to speed up the process, as we will explain later.

There is no better description of the US economic policies during the era of President Reagan in 1980 than that of former President George Bush Sr., who described them as Voodoo Economics, which did not succeed in correcting the trade balance in favor of the USA during the subsequent decades (8).

In his book, *Zombie Economics: How Dead Ideas Still Walk Among Us*, published in 2010, the Australian economist John Quiggin, described the economic concepts that failed to predict three major economic disasters during the past 100 years, from the Great Depression to the 2008-2009 crisis, as "zombies" (9). This rationale was supported by the US economist Paul Krugman in his book *Arguing with Zombies: Economics, Politics, and the Fight for a Better Future,* published in 2021 (10).

In brief, the world is really in need more than ever for new theories that accommodate the changes in the global economy and introduce the concept of innovation and imitation for a deeper understanding of international trade, foreign investment, currency, and monetary policy because the world is in the throes of a hidden economic crisis that may erupt like a destructive volcano at any moment leaving behind a lot of damage that may take more than one generation to rectify.

# References


1. D. Ricardo, *On the Principles of Political Economy, and Taxation.* (John Murray, 1821).
2. R. Vernon, International Investment and International Trade in the Product Cycle. *The Quarterly Journal of Economics* **80**, 190-207 (1966).
3. B. Ohlin, *Interregional and International Trade.* (Harvard University Press, Cambridge, 1933).
4. P. Krugman, A Model of Innovation, Technology Transfer, and the World Distribution of Income. *Journal of Political Economy* **87**, 253-266 (1979).
5. X. Xu, Rethinking the China-US Balance of Trade: 1990-2005. *iBusiness* **04**, 43-50 (2012).
6. *The People's Republic of China* (2020 https://ustr.gov/countries-regions/china-mongolia-taiwan/peoples-republic-china).
7. United States Census Bureau. (United States Census Bureau, 2020).
8. B. Domitrovic, George H. W. Bush's Voodoo Rhetoric. *Forbes.* 2018.
9. J. Quiggin, *Zombie Economics: How Dead Ideas Still Walk Among Us.* (Princeton University Press, 2010).
10. P. Krugman, *Arguing with Zombies: Economics, Politics, and the Fight for a Better Future.* (W. W. Norton, 2020).

# PART TWO

# Copycatting And Imitating Ground-Breaking Innovations Since The Dawn Of History

This chapter overviews a range of innovations throughout human history copied and imitated extensively through ages, and some of them have even remained in circulation so far. The ancient nations and civilizations protected (many of) the profitable innovations they invented. At the same time, their contemporaries endeavored to emulate these innovations by copying, imitating, or even stealing them or their production secrets. Some of these innovations have survived to the present, even if in a new look or different style. Changing silk inscriptions and drawings was not creating a new type of silk, and tailoring *a la mode* silk clothes cannot stand as an innovation of new silk. It has been the same innovative product for thousands of years. This also applies to porcelain, paper, glassware, gear, concrete, dyes, and many other innovations whose contents have not changed over time though their external forms may have been subject to some changes. The introduction of the machine and the industrial revolution only replaced the human element, though partially, and accelerated the manufacturing steps, but the innovative product remained as is. The machine increased production but did not change the original, innovative product.

# CHAPTER FOUR

CHAPTER FOUR

# INNOVATION AND IMITATION: POTTERY AND CHINESE PORCELAIN

Clay and pottery industry has been known since the time immemorial, but man discovered porcelain industry only 1500-4000 years ago reportedly in China during the Shang Dynasty (1046-1600 BC) (1). Historians differ, as some indicated porcelain was invented during the Han Dynasty (220 BC-206 CE), while others argue for a more recent date in the Tang Dynasty (618-907 CE) (2). Recent excavations have revealed porcelain dating back to the Xia Dynasty (1600-2070 BC) (3). Although pottery and clay industry seems similar to that of porcelain, the latter requires a type of white clay soil that contains kaolin along with exposing the material to a temperature of up to 1400 degrees Celsius. In 1600 BC, the Chinese began inventing kilns/ furnaces with a temperature exceeding

1,000 degrees Celsius making some historians assume that pottery was made during (or before) the Shang Dynasty due to the availability of large high-temperature kilns, one of the most important steps in porcelain manufacture (1, 3). Some ancient kilns 16 meters deep that allowed porcelain makers to produce large quantities in less time were recently discovered in China featuring an advanced engineering design, including various levels, holes to intake flame heat while avoiding direct combustion of the raw material, and adequate ventilation to prevent explosion. Moreover, some of these kilns were connected to transfer heat from one to another and without venting except at the seventh or eighth kiln.

Since porcelain had a vitreous luster, the nations that imported it did not think it was made of clay. Some thought its manufacture and luster involved fisheyes and eggshells. The Chinese kept this trade secret for more than two thousand years (assuming it was invented during the Shang Dynasty, or a thousand years if invented during the Tang Dynasty). In the sixteenth and seventeenth centuries, this compound was known in Europe as "white gold." Kings and emperors queued to possess it. Augustus II, King of Poland and parts of Germany, made a palace with porcelain roofs and called it "The Japanese Palace," as detailed later in this chapter (4, 5).

China produced porcelain and exported it to all parts of the world through well-known trade routes. The City of Jingdezhen, southeast China, was associated with the porcelain industry. This city was chosen for its distinguished logistical location that served the porcelain industry. The white clay containing kaolin was in the mountains overlooking the city, while the woods of the nearby forests were used as fuel for kilns. The city was also on a river, a practical means for transportation. Staying for a period in this city, a Chinese traveler reported in 1576 CE, "*How can anyone sleep while the earth trembles with tens of thousands of workers?*" extracting clay. A French who visited a factory in this city reported that making a piece of the porcelain passes through the hands of 72 workers and artisans (6-8).

While the secret of this innovation was kept confidential, the manufacturing process was complicated and required lengthy steps and technical skills. Many attempts to imitate this innovation failed over the ages until the past three centuries. This increased the importance and rarity of porcelain and distinguished it even from silk, violet dye, also known as Tyrian purple, and papermaking, the manufacturing secrets of each were known and reproduced in less time after their discovery.

When the rule of the Arabs in Andalusia ended in the fifteenth and sixteenth centuries, the Portuguese controlled the trade routes in the East while the Spanish controlled them in the West as per the Tordesillas Treaty in 1494 CE after the Pope in the Vatican adopted it. The Spaniards brought in vast quantities of gold and silver from the mines of the Americas. The amount of silver that flowed into Europe (estimated at 3,915,000 tons) doubled the price of bread nearly six times in 150 years (9).

The financial solvency and purchasing power of the Spaniards and the Portuguese rose dramatically, which increased the demand in domestic markets for luxury goods. King Philips II, King of Spain and Portugal during the Union period since 1581 CE, owned more than 3,000 pieces of Chinese porcelain before his death in 1598 CE. Due to the high demand for porcelain, the Dutch East India Company devoted a part of its activity to importing and selling Chinese porcelain. It imported more than three million pieces to Europe from 1604 to 1656 CE. Before this date, King Manuel I (1495-1521 CE), King of Portugal, asked Vasco de Gama to bring India's two most important items, namely spices and porcelain, as the Portuguese believed that porcelain was made in India (4, 5, 10).

In 1983 CE, the British archaeologist Michael Hatcher discovered a Dutch ship called Nanking Cargo sunk by the Portuguese in 1640 CE on its way to Europe containing a large amount of Chinese porcelain, thus reflecting the extent of commercial activity and the high demand for porcelain in Europe at that time (11).

In 1614 CE, Claes Janson Sytmars obtained a patent in Holland entitled "Porcelain Like Imported" for making porcelain. Still, the product was not porcelain but rather lower quality types of ceramics. However, the Dutch's Delftware products in the City of Delft were an imitation of ceramics and created a commercial market for exports to neighboring European countries. Similar ceramic products were previously produced in Italy as part of the attempts to copy and imitate Chinese porcelain (12, 13).

Venice controlled the Silk Road leading to Europe when the Byzantine State collapsed after the Fourth Crusade in 1202-1204 CE. Within 200 years, Venice became one of the most important global trade centers, and its merchants roamed the world in search of exotic and luxury goods to import and then re-export and sell to the rest of Europe (14). When Italians, in general, noted that porcelain became one of the most popular commodities among European nobles and kings, they tried to produce and imitate Chinese porcelain in the fifteenth and sixteenth centuries. Although they failed to access the secret of this industry, they produced ceramics of less quality than porcelain in Faenza under the patronage of Duke Medici, ruler of Florence. This product, popularly known as "Faenza" in Europe, brought an economic renaissance to the Italian cities that produced it. However, it never reduced the consumption of high-quality Chinese porcelain, let alone replacing it (15, 16).

The French imitated this Italian product and marketed it as Majolica ceramics. Imitation attempts continued to develop regarding drawings when the Roman and Greek drawings and vestiges of the Catholic civilization were introduced to ceramic pots and artifacts, which became popular in Europe for decoration and display (17, 18).

In 1688 CE, the French sent ten missionaries to China, one of whom managed to steal the industrial secret of the Chinese innovation after it had been a secret for more than a thousand years. In 1700 CE, François Xavier d'Entrecolles went to the City of Jingdezhen. Twelve years later, he sent a letter with the recipe to the Patriarch of the French Church, who replied that the information was insufficient and not detailed enough to produce the same luster of Chinese porcelain. After further research and investigation, d'Entrecolles prepared a more accurate description of the porcelain production stages in 1722 CE, along with soil samples and plans for the kilns used. In other words, stealing the trade secret of porcelain took more than 22 years after visiting the porcelain factories in Jingdezhen (19).

Despite the relentless attempts over nearly 200 years to steal the secret of porcelain manufacture and imitate its production in Europe, the Italians, the French, and the Dutch failed. However, the King of Poland was lucky after the writings of the French missionary on producing Chinese porcelain were translated and circulated in Europe.

A 19-year-old young man named Johan Friedrick Bottger (1682-1719 CE) claimed he knew alchemy and could produce gold. The soldiers of King Augustus II, who ruled Poland for two terms from 1697 to 1733 CE, kidnapped him. The King provided Bottger with everything he needed in terms of tools and assistants and allowed him to stay in the royal palace next to him on the condition he produced gold. When time passed with no result, Tschirnhaus, another alchemist at the court 30 years older than Bottger, convinced him to turn to white gold (porcelain). Both tried everything that fell into their hands, including eggshells, snakeskin, chalk, and others to no avail. In 1710 CE, it was successfully produced in the German city of Meissen, which was under the rule of the King of Poland. The production of porcelain began for all of Europe. That innovation (or imitation, to be more precise) generated enormous profits for the King. The State exclusively owned the factories as a tributary of the national economy. Thus, Augustus II built the Japanese Palace with decorated ceilings in white gold (porcelain) (20, 21).

The Meissen factory manufactured 1,200 pieces of porcelain in 1711 CE. Although the German porcelain spread as a commodity in the European markets, Augustus II kept the recipe a secret in Meissen. In this way, he increased his economic returns due to producing a product at a price lower than that of Chinese porcelain in European markets. This was due to porcelain's high land and sea shipping cost; it took at least a year and a half from production to reach European markets. As a result, sales of porcelain factories in Meissen increased from 9,696 thalers [a silver coin weighing 16,704 grams and of purity of 93.75 percent] to more than 82,000 thalers in 1742 CE, i.e., more than seven times the profits in nearly thirty years (20, 21).

After the successes and economic impact of porcelain production in Meissen, Frederick II, known as Frederick the Great (1740-1786 CE), King of Prussia, tried to establish a national factory and pumped 6,000 thalers to that end in 1740 CE for six years without success. However, after that, he managed to establish The Royal Prussian Factory with the aid of Wilhelm Wegely after promising him a fifty-year monopoly on the sale of porcelain in the Kingdom of Prussia. It is worth noting that King Frederick II paid 1.84 million thalers to copy and imitate the silk industry but failed. However, he finally succeeded in producing porcelain. He even used his military power to impose porcelain sale contracts on the Ottomans and prevent European competitors from doing so. This policy ensured the marketing of porcelain factories' products abroad and prevented competition (22, 23).

In 1744-1745 CE, the Kingdom of Austria established the Vienna Porcelain Factory in Vienna, which faced six years of continuous failure before manufacturing porcelain. After that, governmental porcelain factories owned by kings and nobles spread in Europe. This period also witnessed attempts by individual "entrepreneurs," especially in Holland and Britain, where the right of monopoly was granted only by patents. In Britain, a potter named Josiah Wedgwood produced porcelain in 1759 CE in Staffordshire and reached the creamware porcelain, then black basalt and blue jasperware porcelain. He decided to gift Britain's Queen Charlotte (1744-1818 CE) a set of creamware porcelain in 1765 CE. Although the product was an imitation, the drawings, designs, colors, and engravings were modern, and the quality of artistry increased the popularity of the commod ity. In 1770 CE, the Russian Empress Catherine II (1762-1796 CE) commissioned Wedgwood to produce a dinner and jewelry set for her. This prompted the Russian emperor later to ask Wedgwood to set up a factory in Russia for creamware porcelain and promised him a grand prize. This conveyed the secret of the porcelain industry to Russia (24-27).

At the time, the Roman Empire was disintegrated into around 300-500 states. Every king and duke tried to produce white gold (porcelain) for himself to get rich, and they succeeded after many attempts. Within a century, porcelain was produced in almost all European countries. This, in turn, declined the import of Chinese porcelain in the nineteenth century. Europe exported porcelain to the East as well (5, 28). The porcelain imitation process shows how imitation eclipsed the benefit of innovation to the innovator who incurred a substantial economic loss while enriching the imitator and the copyist. However, imitation made porcelain available to the middle class and then to the public after it was only a privilege to the nobles, dukes, and kings in Europe. Imitation made the product more abundant, less expensive, and available in more markets. The benefit was generated by technology diffusion for various parties other than the innovator.

# References

1. J. Needham, R. Kerr, *Science and civilisation in China. Vol. 5, Part 12, Vol. 5, Part 12.* (Cambridge University Press, Cambridge; New York, 2004).

2. W. Li *et al.*, A Landmark in the History of Chinese Ceramics: The Invention of Blue-and-white Porcelain in the Tang Dynasty (618–907 A.D.). *STAR: Science & Technology of Archaeological Research* **3**, 358-365 (2017).

3. Y. Li, B. Zhang, H. Cheng, J. Zheng, The Earliest Chinese Proto-Porcelain Excavated from Kiln Sites: An Elemental Analysis. *PLOS ONE* **10**, e0139970 (2015).

4. S. L. Marchand, *Porcelain: A History from the Heart of Europe.* (Princeton University Press, 2020).

5. R. Finlay, *The Pilgrim Art: Cultures of Porcelain in World History.* (University of California Press, 2010).

6. E. Huang, UC San Diego., San Diego, CA, USA (2008).

7. J. Wu *et al.*, Early development of Jingdezhen ceramic glazes. *Archaeometry* **62**, 550-562 (2020).

8. M. Gillette, Copying, Counterfeiting, and Capitalism in Contemporary China: Jingdezhen's Porcelain Industry. *Modern China* **36**, 367-403 (2010).

9. T. Waisberg, The Treaty of Tordesillas and the (re)Invention of International Law in the Age of Discovery. *Meridiano 47. Journal of Global Studies* **18**, (2017).

10. R. Finlay, The Pilgrim Art: The Culture of Porcelain in World History. *Journal of World History* **9**, 141-187 (1998).

11. M. Berg, In Pursuit of Luxury: Global History and British Consumer Goods in the Eighteenth Century. *Past & Present*, 85-142 (2004).

12. T. Volker, *Porcelain and the Dutch East India Company: As Recorded in the Dagh-Registers of Batavia Castle, Those of Hirado and Deshima and Other Contemporary Papers ; 1602-1682.* (E. J. Brill, 1954).

13. S. Ostkamp. (2014).

14. T. F. Madden, The Venetian Version of the Fourth Crusade: Memory and the Conquest of Constantinople in Medieval Venice. *Speculum* **87**, 311-344 (2012).

15. P. Colomban, V. Milande, H. Lucas, On-site Raman analysis of Medici porcelain. *Journal of Raman Spectroscopy* **35**, 68-72 (2004).

16. C. Le Corbeiller, A Medici Porcelain Pilgrim Flask. *The J. Paul Getty Museum Journal* **16**, 119-126 (1988).

17. F. O. Waage, Preliminary Report on the Medieval Pottery from Corinth: I. The Prototype of the Archaic Italian Majolica. *Hesperia: The Journal of the American School of Classical Studies at Athens* **3**, 129-139 (1934).

18. W. M. Milliken, Italian Majolica. *The Bulletin of the Cleveland Museum of Art* **31**, 7-15 (1944).

19. H. Wu, The Translations of François-Xavier Dentrecolles (1664-1741), Missionary in China: Locality and the Circulation of Knowledge. *Extrême-Orient Extrême-Occident* **36**, 49-80 (2013).

20. P. Colomban, V. Milande, On-site Raman analysis of the earliest known Meissen porcelain and stoneware. *Journal of Raman Spectroscopy* **37**, 606-613 (2006).

21. K. DOMONEY, A. J. SHORTLAND, S. KUHN, CHARACTERIZATION OF 18TH-CENTURY MEISSEN PORCELAIN USING SEM–EDS*. *Archaeometry* **54**, 454-474 (2012).

22. W. O. Henderson, The Rise of the Berlin Silk and Porcelain Industries. *Business History* **1**, 84-98 (1959).

23. W. O. Henderson, *Studies in the Economic Policy of Frederick the Great.* (Taylor & Francis Group, 1963).

24. N. McKendrick, Josiah Wedgwood: An Eighteenth-Century Entrepreneur in Salesmanship and Marketing Techniques. *The Economic History Review* **12**, 408-433 (1960).

25. N. McKendrick, Josiah Wedgwood and Factory Discipline. *The Historical Journal* **4**, 30-55 (1961).

26. A. H. Church, *Josiah Wedgwood, Master-potter*. (Seeley, 1894).

27. B. Dolan, *Josiah Wedgwood: Entrepreneur to the Enlightenment*. (HarperCollins Publishers Limited, 2008).

28. G. Blake-Roberts, A. Rawsthorn, I. Luna, *Wedgwood: A Story of Creation & Innovation*. (Rizzoli International Publications, Incorporated, 2017).

# CHAPTER FIVE

CHAPTER FIVE

# INNOVATION AND IMITATION: CHINESE SILK, PAPER, AND EGYPTIAN PAPYRUS

Silk was discovered by chance about 4,500 years ago, when the Chinese Empress Xi Ling-shi, known as "Leizu," wife of the Chinese Emperor Huangdi, nicknamed the Yellow King due to his control over the cities and kingdoms located on the banks of the Yellow River (2587-2697 BC), was wandering in the gardens of the palace and noticed some worms eating the leaves of mulberry trees. She ordered the servants to collect the larvae of this worm and put them in a basin of boiling water. Noticing thin threads loosening from the larva's shell, she ordered the servants to collect a quantity of these threads and turn them into a dress for her. The dress's fabric was very soft, resistant, and highly elastic. It took a few generations for the silk industry to spread in China. Then one of the kings of China ordered that the secret be kept confidential and remain in China. The silk factories were erected in the villages famous for the abundance of mulberry trees. The invention of the silk industry remained a secret with the Chinese for more than two thousand years, since 2640 BC. As a product or a commodity, it spread worldwide, and kings, ministers, princes, and chieftains all over the world wore it. Women were fond of it. However, it was expensive; its price in Rome before Christ was equivalent to its weight in gold. It was used as a significant commodity in commercial exchanges even before the official inauguration of the Silk Road in 136 BC during the reign of the Chinese Emperor Wu Di (1, 2).

During the reign of the Byzantine Emperor Justinian I (527-565 CE), a dispute between the Persians and the Byzantines in the sixth century CE cut off the silk trade route and obstructed trade convoys between Rome and China; this doubled the silk prices. When two priests who traveled to China told the emperor that silk was produced from the silkworm, he sent them on a mission to China to steal these worms and bring them to Constantinople. However, this was challenging. After years, they managed to bring the larvae eggs. There are many accounts about this. One of them indicated that the priests put the larvae in the branches of the bamboo trees and carried them to Constantinople; another indicates they carried them with white mulberry branches (3-5).

The emperor directly established laboratories for the reproduction of larval eggs and then began to manufacture silk under State protection and enacted laws to monopolize the silk market and prevent the leakage of the newly acquired secret of this innovation (or imitation, so to speak). This industry enriched all the citizens of the Byzantine Empire, regardless of their ranks, as Lopez indicated in 1945 CE. Although Chinese silk was of a higher quality and the local production capacity in the Byzantine Empire was limited and did not even cover the local consumption, the transfer and imitation of this industrial technology were of great economic benefit (3-5).

Although it is challenging to set an exact date for the invention of paper, some sources have indicated that it was produced in ancient Egypt about 5,000 years ago (6, 7). Since it is believed that man did not know writing before that, he knew how to draw. Discovered cave paintings date back more than 13,000 years (8). In the twentieth century, 40,000 pieces of inscribed wood and bamboo and 150,000 pieces of bone carved with drawings and symbols were discovered. Some ruins found later in the Indus region of Tapa (in today's Peru) included carvings dating back to 2100 BC and carvings on clay tablets of the Assyrians in Mesopotamia.

Paper is believed to be humanity's greatest invention. In the period between paper (or papyrus) invention and carving on wood, clay tablets, bones, cave walls, and mountains, writing on animal hides was an intermediate and practical stage until the appearance of paper. Parchment, a material made from the hides of sheep, goats, calves, and other animals, was used for writing. Its resistance, practicality, foldability, and lightweight distinguished it from wood, bone, and clay or rock panels. An advanced type of leather called vellum, a parchment extracted from calves, emerged, but its quality was higher. It was called parch because it was a white sheet of laminated leather. The name was after the hides lamination process (9, 10).

The ancient Egyptians made paper from the stem of the papyrus plant, which grew in the ancient Nile Delta. However, this plant lived only in wet areas and close to rivers. When Alexander the Great (336-323 BC) decided to establish the Great Library in Alexandria (the city he founded upon his arrival in Egypt), he placed the library near the paper mill where these plants were abundant. Three centuries later, the books made of papyrus exceeded 700,000. The imitation and manufacture of papyrus moved from the ancient Egyptians to the Greeks, the

Romans, then the Arabs, and so it spread. However, the industry remained limited due to the lack of enough papyrus plants and the special conditions required for their growth near swamps and rivers. The Romans also used vinegar and flour with papyrus to cover writing to remain longer, not be easily erased, and resist the paper weevil (6, 7, 11).

When the Spanish sailor Hernán Cortés (1485-1547 CE) arrived in the Americas in the sixteenth century, he found that the Aztec Empire (1428-1521 CE) (the indigenous people of an area in today's Mexico) knew how to make and use paper (12).

The paper was reinvented in China during the Han Dynasty (202 BC-220 CE) from materials other than the papyrus plant. A fragment of a paper dating to 141-179 BC made from mulberry bark was found in Gansu State, the City of Fangmatan. Over the centuries, other trees (besides berries) were used, such as hemp, bamboo, wheat straw, sandalwood, and cedar (13).

In the sixth century (about 610 CE), the Buddhist priest Dam Jing transferred the papermaking mechanism to Japan. Paper had a variety of uses, including wrapping antiques, cloth ornamentation, correspondence, books, or even making banknotes (14). It is worth noting that banknotes began in 1023 CE (or 960 CE in the era of Thai-tsu) as a security certificate against depositing gold and silver in the treasuries of the Song Dynasty; it became a stand-alone currency (not against deposit) in 1120 CE and the right to issue it was monopolized by the government of the Song Dynasty. This is one of the later benefits of paper innovation that we see today (15).

The secret of papermaking remained in the Far East until the Battle of Talas River in 751 CE between the Islamic Abbasid State and the Tang Dynasty (618-907 CE). The Abbasids won, and two of the captives were papermakers. They were transferred to the City of Samarkand, and a paper factory was built there, then the industry moved to Baghdad, and the use of paper spread widely as a commodity in most parts of the world at that time. The Arabs used flax and grass to substitute tree bark used in China (16). In 1221 CE, the Roman Emperor Frederick II restricted paper to government documents to protect the local product, arguing that paper is more perishable than leather (17).

In the eleventh century, the paper industry was introduced to Europe through the Arab state in Andalusia. It was reported that a French priest visited Andalusia in the twelfth century and noted that Arabs and Jews there used paper instead of leather parchment to write their religious texts. Although the Arabs tried to sell and market the paper commodity to Europeans through trade channels, they rejected that for hundreds of years as they insisted on using parchment, which only priests and rulers needed (12).

In 1441 CE in Germany, a German blacksmith named Johannes Gruenberg invented the paper printing press, which revolutionized printing. The process of preparing one copy of the Bible needed the hides of 250 sheep (18). In 1797 CE, Louis Nicolas Robert manufactured a papermaking machine in France. An English sculptor named Henry Fourdrinier obtained a patent in 1806 CE to manufacture paper, known later as the Fourdrinier machine (19). Russian Tsar Nicholas asked Henry to make two machines in Peterhof, Russia, for £7,000 (12).

Thus, the paper industry was copied and imitated with a paper pressing and laminating machine from Germany to France and then Britain and Russia. Innovation has been widely imitated. Imitation was the reason for the spread of papermaking technology. The growth of this industry continued in Europe until the British annual production of paper reached nearly 100,000 tons and then exceeded half a million tons in 1900 CE.

# References


1. W. S. Kilborne, Silk. *Industrial & Engineering Chemistry* **44**, 2166-2168 (1952).
2. R. Rakshith, C. Om Prakash, A Study on Working Capital Management at Karnataka Silk Industries Corporation Ltd.,. *Bengaluru*, 1-93 (2019).
3. H. B. Feltham, Justinian and the International Silk Trade. *Sino-Platonic Papers* **194**, (2009).
4. J. Freewalt, "Justinian and China: Connections between the Byzantine Empire and China during the Reign of Justinian I," *Late Antiquity and Byzantium* (American Military University 2015).
5. R. S. Lopez, Silk Industry in the Byzantine Empire. *Speculum* **20**, 1-42 (1945).
6. E. G. Turner, *Greek Papyri: An Introduction*. (Princeton University Press, 2015).
7. J. A. Harrell, V. M. Brown, The World's Oldest Surviving Geological Map: The 1150 B.C. Turin Papyrus from Egypt. *The Journal of Geology* **100**, 3-18 (1992).
8. E. Callaway, Skeleton plundered from Mexican cave was one of the Americas' oldest. *Nature* **549**, 14-15 (2017).
9. M. L. Ryder, Parchment—its history, manufacture and composition. *Journal of the Society of Archivists* **2**, 391-399 (1964).
10. M. Fourneau *et al.*, Histological study of sheep skin transformation during the recreation of historical parchment manufacture. *Heritage Science* **8**, 78 (2020).
11. C. Römer, in *A Companion to the History of the Book*. (2019), pp. 191-202.
12. M. Kurlansky, *Paper: Paging Through History*. (W.W. Norton, 2017).

13. P. Shenoy, P. Aithal, A study on history of paper and possible paper free world. *International Journal of Management, IT and Engineering* **6**, 337-355 (2016).

14. H. Won, Test of the hybrid origin of Broussonetia × kazinoki (Moraceae) in Korea using molecular markers. *Korean J. Pl. Taxon* **49**, 282-293 (2019).

15. J. Pickering, The History of Paper Money in China. *Journal of the American Oriental Society* **1**, 136-142 (1844).

16. S. A. Gunaratne, Paper, Printing and the Printing Press: A Horizontally Integrative Macrohistory Analysis. *Gazette (Leiden, Netherlands)* **63**, 459-479 (2001).

17. R. Goldsborough, In Celebration of Paper. *Tech Directions* **72**, 12 (2012).

18. F. Rees, *Johannes Gutenberg: Inventor of the Printing Press*. (Capstone, 2005).

19. M. Blackman, From papermaking to aircraft navigation: Significant patent and inventor anniversaries in 2006. *World Patent Information* **28**, 336-343 (2006).

# CHAPTER SIX

## CHAPTER SIX

# INNOVATION AND IMITATION: GREEK FIRE, ROMAN CONCRETE, AND PHOENICIAN TYRIAN PURPLE

Kallinikos, or Callinicus, invented the Greek fire in Baalbek (located in today's Lebanon) in 672 CE. It consisted of a combustible compound emitted by a flame-throwing trumpet-like weapon. The trumpets contained a tank filled with a liquid substance composed of a mixture of materials. After the Byzantines adopted this innovation during the reign of Emperor Constantine IV (668-685 CE), the naval warship was supplied with Greek fire to burn enemy ships and castle towers. The Byzantines also designed mobile chariots that could be pushed during military campaigns with the same weapon. Because of this weapon, Emperor Leo III the Isaurian managed to stop the military advance in 717-718 CE, led by Maslama bin Abd al-Malik. The Byzantines kept the secret of this invention for generations. Emperor Constantine VII (913-959 CE) told his son and successor to the throne Romanos II (959-963 CE), to be careful so that the barbarians do not know this secret. But fate meant otherwise when the Islamic kingdoms overcame the Byzantines, took the secret, and reproduced the weapon (1-3).

Around 1200 BC, the Phoenicians in the Levant invented a violet dye called Tyrian Purple after the name of the ancient Tyre. This dye was extracted from the sticky mucus membranes of a violet murex shellfish from the bottom of the Mediterranean Sea. This murex trunculus medium-sized shellfish of the genus truncularіopsis contained bromide compounds, such as 6-6-Dibromoindingo, which provided the violet color. Producing this dye was labor-extensive and required tens of thousands of shellfish, which made a small weight thereof worth a fortune. It took 12,000 shells to produce 1.4 grams of dye (4, 5).

After the Romans' conquest of the Levant, they copied this innovation. Later, it was used to decorate the clothes of kings and nobles in Rome. In the Roman Empire in the fourth century CE, a law was enacted prohibiting ordinary people from wearing silk dyed completely with this innovative violet dye because it was the exclusive right of only the Roman Emperor. The Roman senators were allowed to put a purple ribbon on the edge of their togas worn while the senate was in session (6).

It also spread in Greece and Byzantium and remained an industrial secret until Constantinople fell during the Fourth Crusade; it spread after that. In the Byzantine era, there were special laboratories to process raw silk brought from China, dyed in purple, and partially re-exported at a high value to China. It was reported that Louis the Pious, King of France, was a brother of one of the emperors of Byzantium. When the quantity of violet-dyed silk clothes was scarce in his country, he wrote to the Emperor of Byzantium (his brother) to send him artisans with the secret of the violet dye. When his brother refused, Louis devoted artisans and scientists to produce such a dye, which they did but at a lower quality (7).

In the second and third centuries BC, the Romans invented a type of concrete resistant to the surrounding conditions, which helped some buildings survive for more than 2000 years. They added the volcanic ash brought from Pozzuoli in Nablus (currently Italy), 100 km south of Rome. It contained siliceous-aluminous, which reacted with calcium hydroxide to form high cementitious cohesion properties (8, 9).

In 30 BC, Marcus Vitruvius Pollio described the mixture process in preparation for construction in Rome and referred to mixing one share of limestone with three shares of pozzuolana, also known as pozzolanic ash, for any construction on land and one share of limestone with two shares of pozzuolana for any construction underwater, to make the structure remain longer (10).

Although the Romans were not the first to use concrete for construction, as Nabatean Arabs south of the Levant during the period dating back to 6000 BC preceded them, followed by the Mesopotamians and then the Pharaohs in Egypt, the Romans used volcanic ash to harden concrete further and prevent cracks in buildings. Such an addition made some buildings stand for more than 2000 years, notably those at sea, saltwater shores, and ports (11). A research team from the University of California at Berkeley has recently discovered that the Romans added seawater to the concrete mixture for additional resistance properties (12).

It is worth noting that the Romans knew how to distinguish different types of volcanic ash and determine their suitability for construction. During the reconstruction of the Island of Crete in the Mediterranean, volcanic ash was available in abundance in the nearby city of Santorini. Thus, the Romans brought the pozzuolana volcanic ash far away because of its quality. With the Roman control over Europe, North Africa, and the Levant region, the Romans transferred this innovation to those regions. There, the technology was copied or imitated and then spread.

An analytical view of the Roman concrete innovation shows that concrete used limestone in construction since 6000 BC, i.e., more than 5,000 years before the Romans knew that innovation. The volcanic ash was discovered and used in Greece in the fifth century BC, about 300 years before the Romans used it. One of the greatest inventions of the Romans was imitating other peoples' innovations and adapting them to match the raw materials available in the Roman Empire's local environment, then using the volcanic ash from volcanic sources near the capital, Rome. This ash granted concrete better properties. In other words, the Romans began to imitate the concrete available in the neighboring countries and then developed the process by improving the types of inputs in the concrete mix. Thus, their innovation followed imitation.

# References


1. A. Roland, Secrecy, Technology, and War: Greek Fire and the Defense of Byzantium, 678-1204. *Technology and Culture* **33**, 655-679 (1992).

2. J. R. Partington, B. S. Hall, *A History of Greek Fire and Gunpowder*. (Johns Hopkins University Press, 1999).

3. A. Mayor, *Greek Fire, Poison Arrows, and Scorpion Bombs: Biological and Chemical Warfare in the Ancient World*. (Overlook Duckworth, 2003).

4. C. J. Cooksey, Tyrian Purple: 6,6'-Dibromoindigo and Related Compounds. *Molecules* **6**, 736-769 (2001).

5. C. Cooksey, Tyrian purple: the first four thousand years. *Sci Prog* **96**, 171-186 (2013).

6. L. B. Jensen, Royal Purple of Tyre. *Journal of Near Eastern Studies* **22**, 104-118 (1963).

7. J. Edmonds, *The Mystery of Imperial Purple Dye*. (John Edmonds, 2000).

8. C. J. Rhodes, Roman Concrete for Durable, Eco-Friendly Construction – Applications for Tidal Power Generation, and Protection against Sea Level Rise. *Science Progress* **101**, 83-91 (2018).

9. C. J. Brandon, R. L. Hohlfelder, M. D. Jackson, J. P. Oleson, *Building for Eternity: the History and Technology of Roman Concrete Engineering in the Sea*. (Oxbow Books, 2014).

10. V. Pollio, J. Gwilt, *The Architecture of Marcus Vitruvius Pollio: In Ten Books*. (Lockwood, 1874).

11. N. J. Delatte, Lessons from Roman Cement and Concrete. *Journal of Professional Issues in Engineering Education and Practice* **127**, 109-115 (2001).

12. A. Witze, Seawater is the secret to long-lasting Roman concrete. *Nature*, (2019).

# CHAPTER SEVEN

CHAPTER SEVEN

# INNOVATION AND IMITATION: PTOLEMAIC GEER, PHARAONIC KOHL, AND SUMERIAN GLASS

Geer (or cogwheel) are one of the greatest innovations, second only to the invention of the wheel and paper, because of their various benefits in lifting heavy weights, construction applications, water transport, irrigation, and military equipment. Although it is difficult to set a date for this innovation, some historians have indicated that cogwheel were invented in Alexandria, Egypt, during the Ptolemaic Kingdom (305-30 BC) in the third century BC. Ctesibius (230-270 BC) wrote a description of gears as later documented by Philo, a Byzantine scholar. Hernon of Alexandria also wrote in 25 BC a description of this innovation. A set of connected bronze cogs/ gears rotating in different directions believed to be used to transfer water for bathing was found in ancient Egypt's temples in 287 BC. The *saqiya* (waterwheel) was also a mechanical irrigation system that used animals, such as oxen, to move a large geer connected to columns and other geers to bring water out of the wells. The *saqiya* is still used today in some villages (1).

Accordingly, geers were invented in Alexandria and then copied or imitated in the cities of Greece and the islands of the Mediterranean between Egypt and Greece. Archimedes (212-287 BC) used them on the Island of Zaragoza. When the Romans took control of Greece in the first and second centuries BC, they copied the geer technology and settled it in Rome first and then in other states, including the Levant and Turkey. Over time, this innovation was transferred to the rest of the world, and the use of geers continues today. The reader may have a watch on his wrist that contains dozens of tiny geers (2).

Many beauty products are innovated every day to suit all tastes. The global market for cosmetic products was estimated in 2018 at US$455.5 billion, with an expected annual growth of more than 5%. The returns of some beauty companies have reached US$30 billion, such as the French company L'Oreal, whose sales exceeded US$31 billion a year (3, 4).

The first cosmetic innovation is kohl, invented in Egypt in the fourth century BC. A quantity of kohl was discovered in the Badarian tombs of an Egyptian civilization that preceded the era of the Egyptian kingdoms, dating back to 3000-5000 years BC. The ancient Egyptians mixed galena stone containing lead sulfide (PbS) and malachite stone containing copper carbonate oxides ($Cu_2(CO_3)(OH)_2$) with a bright green color to produce kohl (5, 6).

These stones, available in the Sinai and Eastern Deserts, were used as raw materials in the production process. Galena was imported from the Arabian Peninsula. The mixing ratio of these materials varied to produce multiple colors limited to shades green, gray, black, and blue. However, the blue appeared later when the lapis lazuli stone was ground and used in kohl preparation. The lapis lazuli stone was brought from Badakhshan Province in Afghanistan. Galena stone was the main component of kohl. A research team analyzed six different samples from ancient Egypt around 2000 BC from the British Museum and found that galena in kohl ranged from 40% to 100% (7).

An economic view of the kohl markets would show that kohl was produced in Egypt, and raw materials were brought from neighboring or distant countries, such as the deserts of the Arabian Peninsula or the Afghan mountains. Then quantities were exported to countries that had trade links with Egypt. The Greeks, then the Romans, began importing the product and imitating the methods of its manufacture. However, kohl was well received in the Persian Empire, where it became an indispensable part of daily cosmetic habits and local traditions. Thus, kohl was sold as a product exported from Egypt, and then it became a product manufactured by the Greeks, Romans, and Persians. Imitation of innovation contributed to this industry's transfer from one place to another.

As for glass, its invention dates to the Mesopotamian civilization in the Harran region of the ancient Kingdom of Sumer in 3500 BC. The Sumerians produced glass by cooling volcanic eruptions forming the so-called obsidian, which is volcanic black glass. Then this knowledge passed to the Phoenicians through trade with Mesopotamian civilizations. The oldest pieces of glass found date back to 2500 BC in northern Syria in Cendebia, a Phoenician province on the Mediterranean. However, the Phoenicians copied the technology, reproduced it in their cities, and used it in trade across the Mediterranean, especially in trade exchanges with Egypt. The oldest pieces of glass were found in Egypt, dating back to 2200 BC (8).

In 1500 BC, Egypt, which succeeded in imitating glass production processes, became the first and primary glass manufacturer globally. Glass was exported from Egypt to all parts of the world. The ancient Egyptians developed Sumerian and Phoenician technology for glass production. They added colors to the glass, a revolutionary step that inaugurated the first golden age of glass industries and made glass part of manufacturing all daily household items, such as pots, bottles, jars, and cups. The Egyptians knew that adding calcium to glass during manufacture would give it a white color while adding lead would render a yellow color; chromium would provide an orange color. Adding copper would give red color (or green depending on the degree of copper oxidation). Manganese gave glass its violet color, nickel brown color, and cobalt blue color. In summary, they reproduced glass in almost all colors, which revolutionized the glass industry in that era (8).

The Greeks and Romans learned the technique of glassmaking from the Egyptians. In the fourth century CE, this industry reached its peak in Rome. When Venice occupied Constantinople, which was famous for thriving glass industry in the thirteenth century and was the primary source of glass, Venice copied and imitated this industry and built the Island of Glass, which was launched in 1292 CE on the Island of Murano, kicking off the second golden age of glassmaking in Venice. Barorier removed impurities from glass and made it completely transparent (as it is today), thus leading Venice to produce clear glass eye lenses. Venice also became a producer and exporter of glass to all Mediterranean countries (9, 10).

In 1590 CE, the Dutch Hans and Zacharias Janssen manufactured the telescope in Middleburg after identifying the magnifying feature of glass. In 1565 CE, Robert Hooke of Britain manufactured the microscope using glass lenses. In 1940 CE, television was produced using glass technology. Glass was also used to produce many measuring instruments, such as the thermometer (11). The market for glass products is still thriving today, as the global glass market reached US$98 billion in 2018 (12).

As an analysis of the history of glass innovation, glass was invented by the Sumerians and then flourished as a trade at the hands of the Phoenicians due to their navigation ability and network of trade channels in the Mediterranean Sea. Then it brought about an industrial revolution in Egypt, which brought it home from the Phoenicians first, developed its means of consumption, and then exported it to all parts of the world. The process of copying and imitating glass innovation continued from one country to another and from one generation to another until the modern era. In the second decade of the twenty-first century, the most advanced countries still own glass factories to meet consumer needs in their domestic and international markets. Today, these developed countries have been the leaders in imitation and copying to which they owe their current production and industrial capabilities.

# References


1. M. J. T. Lewis, Gearing in the ancient world. *Endeavour* **17**, 110-115 (1993).

2. V. Vullo, *Gears: Volume 3: A Concise History*. (Springer International Publishing, 2020).

3. M. Ridder. (Statista.com, 2020).

4. D. B. M. research. (Data Bridge Market research, 2020).

5. P. Wexler, *History of Toxicology and Environmental Health: Toxicology in Antiquity II*. (Elsevier Science, 2014).

6. M. Marsh, *Compacts and Cosmetics: Beauty from Victorian Times to the Present Day*. (Remember When, 2009).

7. A. D. Hardy *et al.*, in *Physical Techniques in the Study of Art, Archaeology and Cultural Heritage,* D. Bradley, D. Creagh, Eds. (Elsevier, 2006), vol. 1, pp. 173-203.

8. S. C. Rasmussen, *How Glass Changed the World: The History and Chemistry of Glass from Antiquity to the 13th Century*. (Springer Berlin Heidelberg, 2012).

9. A. Dorigato, *Murano: Island of Glass*. (Antique Collectors Club Limited, 2013).

10. C. I. Gable, *Murano Magic: Complete Guide to Venetian Glass, Its History and Artists*. (Schiffer Pub., 2004).

11. S. Johnson, *How We Got to Now: Six Innovations That Made the Modern World*. (Penguin Young Readers Group, 2018).

12. Market Research Future, "Flat Glass Market," (Market Research Future, 2021).

PART THREE

# DEBUNKING EUROPEAN INNOVATIONS FROM THE FOURTEENTH TO THE NINETEENTH CENTURIES

# CHAPTER EIGHT

## CHAPTER EIGHT

# THE ERA OF COPYCATTING, IMITATION, AND KNOCKING OFF IN EUROPE

From the fourteenth century to the middle of the nineteenth century, the European innovation concept was confined to imitation and copycatting. Imitation was an innovation. Other nations' innovative products and methods were new to the Europeans, whether the manufacturing process or the outcome. The definition of innovation reflected their lack of knowledge of industries and production, rendering every imitation of the products of the East novel to them, hence an 'innovation.' There was no clear-cut distinction between the imitative and the innovative. Instead, every imitation was an innovation and an invention to them. Europe described all counterfeit products and industries as innovations for hundreds of years. While they were new to Europe, they were innovative and present in the other countries that invented them (1-4).

In his works, Benoît Godin pointed out that imitation was the primary meaning of innovation in European history. The concept of modernity and newness did not fall within the meaning of innovation in history. The process of innovation was a 'purely' imitation process. Copycatting and imitating innovative products was named innovation in Europe (1-4).

If we want to analyze the conditions that led Europeans to describe all copycatting and imitating as innovation, we must go back to two periods in ancient European history. The first is the era of the Greeks and Romans, who developed many technologies and witnessed thriving sciences and industries. However, their civilizations never reached present-day Europe despite their wealth of knowledge. Some historians point out that the Gothic and Germanic tribes obliterated the sciences of the Romans and Greeks. For example, although the glass and concrete industry flourished in Rome, they completely disappeared from Europe in the tenth century CE along with many other techniques. In short, the civilizations in Roman and Greek did not connect to the Middle Age Europe (5-7).

The second is the post-Roman period during which Europe did not possess any advanced industries worthy of transferring via the trade routes and the Silk Road. The Europeans were invading each other and selling captives as slaves to foreign merchants on the Silk Road against the products of the East. Slaves were the only thriving commodity in five centuries, followed by natural products, such as fur, honey, and iron (8).

The lack of European products weakened trade exchange with other nations via ancient trade routes. Only a few goods from the East reached the hands of only kings and nobles in Europe. Accordingly, everything from the East was new to the Europeans (8, 9).

Thus, every imitation in the subsequent period was an innovation for the Europeans and a source of pride worthy of con cessions and monopoly rights, as detailed in the next chapter. If we want to set a start date for this European shift to the imitation of the innovations of the East, two separate main events should come to mind. The first was in Eastern Europe, led by the Italian merchant republics, notably Venice. The second was in Western Europe, led by the Portuguese and the Spaniards, immediately after forcing the Arabs out of Andalusia.

To elaborate, after the Fourth Crusade (1202-1204 CE), the Venetians, with the help of the European Crusader armies heading east to Jerusalem, controlled Constantinople (Istanbul today) and occupied it directly, despite it being a Christian empire (they also occupied independent Christian coastal cities, such as Zara, which is located in today's Croatia). The Pope in Rome deplored these acts of the Crusader armies. The Venetians' occupation of Constantinople allowed them to control the entire Adriatic Sea. Since Greece was under the rule of the Byzantine Empire, it was also subject to the Venetian influence along with hundreds of islands and ports in the eastern Mediterranean. This geographical location allowed Venice to open a channel between East and West; everything that passed to Europe via the Silk Road passed through the Venetian spheres of influence. This stage took nearly two hundred years to develop after the Fourth Crusade until its peak at the end of the fourth century. This distinguished geographical location of the Venetians and their skill as sea merchants allowed them not only to get acquainted with the latest innovations and goods brought from the East, but also the market demand on each commodity passing through their areas of influence. They also had trade channels and a communication network linking East to West (10).

That period witnessed attempts to access the industrial secrets of the most sought-after products via the Silk Road. Venetian merchants attempted to steal the industrial secrets to reproduce these products in Venice and accumulate wealth. The Venetians knew the size of the trading markets of these commodities, which prompted the merchants, supported by the government of the merchant republic, to roam here and there and search for the places of production of the most popular commodities. They reached Egypt and the Levant first, then Russia and China. In his writings, the traveler Marco Paolo summarized his experience and that of his father and uncle to find out the secret of an industrial process to transfer it to Venice to get rich (11).

Porcelain, silk, glass, and dyes were not manufactured in the Venetian spheres of influence or Europe. Trade secrets of different nations were stolen, and counterfeit products were reproduced, albeit with lower quality, in the cities of Venice. The Venetian Murano Island was turned into a hub for glassmaking and was called Glass Island. Regulations and laws were enacted to protect innovations (or imitations, so to speak) newly introduced to the Venetians though some of them had existed for hundreds or even thousands of years. We will discuss in an upcoming chapter what happened to Venice and then the rest of Europe.

The second event that kicked off the European imitation era was the Portuguese search for new trade routes to set themselves free from the commercial influence of Venice, Genoa, Pisa, and Florence. These Italian cities-cum-republics controlled the trade routes within Europe. The commodities had to pass through the Italian republics to reach the Spaniards and the Portuguese in the far West of Europe, far away from the trade route. The result was a delay and a higher cost. Moreover, the bloody conflict and religious disagreement between the Arabs in the Maghreb and the Spanish and the Portuguese, especially after their exodus from Andalusia, worsened the situation. Accordingly, the goods produced in the East and Egypt, in particular, did not reach Andalusia through the Maghreb after the Arabs were expelled from it (12).

In light of the scarce resources and poor relations with neighboring countries, the Portuguese dreamt of a road to India to reach the source of the Silk Road without the need for the Italian republics and Venice in particular, which increased prices or to communicate with the Arabs, their religious enemies. The Portuguese believed that porcelain and silk originated from India; they did not know they were produced in China and reached them via the Silk Road through India (12).

In general, this book is not a history of the Portuguese or the Venetians. Still, it is necessary to look at the attempts of the Portuguese kings in the fifteenth and sixteenth centuries, starting with Prince Henry the Navigator (1394-1460 CE), his nephew King John II (1455-1495 CE), and King Manuel (1495-1521 CE). Those efforts culminated in discovering a maritime route that allowed the Portuguese to circumvent Africa through the Cape of Good Hope reached by the sailor Bartolomeu Dias with two ships in 1488 CE. Then followed the crossing of the Arabian Sea and the Indian Ocean to reach India by Vasco da Gama in 1498 CE and the visit of the Kingdom of Zamorin in Calicut. The sailor Afonso de Albuquerque (1510-1515 CE) reached Indochina and the Spice Islands in Indonesia (12-14).

The Portuguese brought some unknown goods to Portugal through the sea trade routes. As for the known items, they were at a much lower price. The Portuguese markets prospered thanks to the availability of innovative oriental goods sold to the rest of Europe. The need for the Italian republics of Genoa and Venice decreased. The availability of these new goods, including dyes and ceramics, prompted their imitation in Portugal, but their attempts were less than those of the Venetians (12, 13).

In the same period, the Spaniards, after employing Portuguese sailors of Italian origin (Genoa), such as Christopher Columbus, managed to reach the Caribbean islands on three ships in 1492 CE. These islands were at the outset of the route of the Americas. Similarly, the new goods brought by the Spanish sailors from the Americas opened new markets, which subsequently stimulated their imitation, reproduction, or even re-cultivation, such as potatoes and corn[4] (15).

4 Potatoes, corn, cocoa, and tobacco, products of consumer value to the indigenous peoples of the Americas, were unknown outside the Americas. After the arrival of the Spaniards to these areas, they transferred goods first and then the consumer culture of these goods to Europe and the rest of the world through known trade networks and routes.

The European imitation and copycatting era began in the fourteenth century and continued to the early nineteenth century. These two events [(1) the Italian republics' control over the trade routes on the Silk Road leading to Europe, and (2) the discovery of the new sea routes by the Spanish and Portuguese to import and transport goods] increased the abundance of innovative products from the East, which Europeans did not know before as they reached only the kings and nobles in Europe rather than the ordinary people due to their scarcity, high prices, and difficulty in obtaining them. Imitation and copycatting began and developed gradually up to the concept of real innovation at the end of the nineteenth century.

The availability of innovative goods through the two events referred to above was the main reason that triggered the era of imitation in Europe during which any imitation was dubbed as an innovation. This concept lasted for nearly five hundred years (2, 3) during which the Europeans knew quite well how backward they were in terms of cultural backwardness and how weak in terms of productive and industrial capabilities compared to the nations in the East. Such knowledge prompted them to break away from religion, abandon it, and accuse it of being the main cause of retrogressivness. The Protestant doctrine emerged, the the Catholic doctrine declined, and secularists who believed in science and weak religious guidance, which led to their industrial backwardness, surfaced. All this was not due to the development of science and knowledge of the Europeans, nor their sudden metamorphosis into innovators, but rather because they recognized their low technical capabilities in industry and innovation compared to those of the civilized peoples of the East (16, 17). After hundreds of years of religious, political, and social upheaval, they embarked on the era of imitation followed by hundreds of years to authentic innovation. It was a journey of five hundred years that began with a severe civilizational shock, followed by stages of changes that paved the way for the launch of the industrial revolution in Europe from the end of the eighteenth century to the middle of the nineteenth century.

This chapter does not claim there was no innovation in Europe from the fourteenth to the nineteenth century. We undertake a second reading of history after analyzing the patents and patenting systems of that era (in the following four chapters) and calling things as they really were, not as what was common at the time. The public order, economic situation, and social psychology of the European societies back then yearned for the products of the East, its elaborate industries, and its exquisite inscriptions and decorations. Although these were innovations produced hundreds and even thousands of years ago, Europe considered them as modern innovations and the reason for the economic renaissance of the nations of the East. The designs were exquisite and stylish for the aristocratic European consumer. These goods did not reach the ordinary Europeans (except rarely), as they were the privilege for the elite and the velvet class.

This chapter (and the subsequent four chapters) is not a negative criticism of Europe, but rather a detailed explanation of innovation and its systems, which helped Europe move from imitation that lasted for five hundred years to innovation in the eighteenth and nineteenth centuries with the spread of steam engine machines and the use of coal as fuel.

Copycatting and imitation may be an effective way that eventually leads to innovation, a strategy that the so-called non-industrial and non-innovative countries may adopt. The European model relied on copycatting and imitation until it reached innovation after about five centuries of trial and error. Through these attempts, institutions were established, systems were developed, regimes were overturned, and new religious and intellectual currents emerged to align with the prevailing innovative (or imitative) thought.

The arrival of the advanced goods from the East through Venice first through the Silk Road and then through the

Portuguese, caused an intellectual revolution that changed beliefs, institutions, and kingdoms to simulate the technical and industrial progress of the East. A civilizational shock yielded its fruits after 500 years of restructuring institutions in politics, religion, economy, and industry. Even the individual's mentality changed to keep pace with the East's progress. Because some religious concepts in Europe were strict about change and modernity, new doctrines, beliefs, and intellectual currents emerged to take people out of the familiar religion claimed to be the reason for Europe's backwardness for more than a thousand years. We may not agree with these views, but this was the claim of many after the culture shock. Every imitation of the technologies and industries of the East was called 'innovation'. This concept remained for hundreds of years in Europe. When the Industrial Revolution flourished, the concepts began to get gradually corrected.

# References


1. B. Godin, *The Idea of Technological Innovation: A Brief Alternative History*. (Edward Elgar Publishing, 2020).
2. B. Godin, *Innovation Contested: The Idea of Innovation Over the Centuries*. (Taylor & Francis, 2015).
3. B. Godin, Innovation: the history of a category Working Paper. *Institut national de la recherche scientifique, Centre Urbanisation Culture Société, Montréal* (2008).
4. B. Godin, *The Invention of Technological Innovation: Languages, Discourses and Ideology in Historical Perspective*. (Edward Elgar Publishing, 2019).
5. J. Nasim, *The establishment of Universities in the Middle Ages (in Arabic Language)*. (Bibliotheca Alexandrina, ed. 2nd 1981).
6. A. B. Cobban, *The medieval universities: their development and organization*. (Methuen, London, ed. 1. publ., 1975).
7. B. L. Drake, Changes in North Atlantic Oscillation drove Population Migrations and the Collapse of the Western Roman Empire. *Scientific Reports* **7**, 1227 (2017).
8. P. Frankopan, *The Silk Roads: A New History of the World*. (Bloomsbury Publishing, 2015).
9. M. Berg, *Luxury and Pleasure in Eighteenth-Century Britain*. (OUP Oxford, 2005).
10. R. Crowley, *City of Fortune: How Venice Won and Lost a Naval Empire*. (Faber & Faber, 2011).
11. M. Polo, *Travels of Marco Polo*. (Penguin Publishing Group, 2004).
12. R. Crowley, *Conquerors: How Portugal seized the Indian Ocean and forged the First Global Empire*. (Faber & Faber, 2015).

13. A. R. Disney, *A History of Portugal and the Portuguese Empire: Volume 1, Portugal: From Beginnings to 1807*. (Cambridge University Press, 2009).

14. A. J. R. Russell-Wood, *The Portuguese Empire, 1415-1808: A World on the Move*. (Johns Hopkins University Press, 1998).

15. H. Thomas, *Rivers of Gold: The Rise of the Spanish Empire*. (Penguin Books Limited, 2013).

16. J. Delacroix, F. Nielsen, The Beloved Myth: Protestantism and the Rise of Industrial Capitalism in Nineteenth-Century Europe*. *Social Forces* **80**, 509-553 (2001).

17. P. Harrison, *The Bible, Protestantism, and the Rise of Natural Science*. (Cambridge University Press, 2001).

# CHAPTER NINE

CHAPTER NINE

# ROOTS OF THE EUROPEAN IMITATION AND COPYCATTING METHODOLOGY

Due to the civilizational shock that took place after importing the innovations of the East (as explained in the previous chapter), a set of theories, intellectual methodologies, and economic concepts emerged, and the European countries adopted them one after another. Thus, we must consider some economic concepts adopted by Europe during the era/period of imitation and copycatting based on which (i.e., these economic concepts) it legislated the right to imitate, copycat, and issue patenting systems in its various stages (the following three chapters). Europe had gone through two major phases. The first was the stage of patent concessions, known as Letters of Patents, by the ruler, king, or duke [next chapter]. The second was issuing patents by institutions that regulated the process [11 & 12 chapters].

The influx of innovative goods (innovative for Europeans despite having existed for hundreds or thousands of years in the countries that produced them) drained silver (and gold) from Europe. Some analysts of the time indicated that all the silver brought from the mines of the Americas was sent to the East to bring in its products. "*Europe dried up its silver brought from America [the new colonies].*" Funds were sent for the unparalleled innovative goods in Europe, hence the urgent need to create local industries for the same products (1). The only way to do that was by knocking off and imitating these products (2, 3).

The aesthetic aptitude of the European aristocracy began to improve to relish the ingenuity and creativity of goods and products after the influx of products from the East. Books and volumes were authored to explain methodologies for tasting antiques, clothes, and other goods and how to differentiate between varying aesthetical degrees and types. The French Gabriel Tarde wrote *The Laws of Imitation* to explain the art of imitation of the culture of the East, the impact of imitation on societies, and how an innovative product would circulate and become common to a society. In 1753 CE, the Englishman William Hogarth wrote *The Analysis of Beauty* on the methods of analyzing aesthetics in goods and products, in terms of their relevance, simplicity, complexity, diversity, and quality. Alexander Gerard also wrote *Essay on Taste* to show the harmony and modernity in goods and products (4-6).

An example of the volume of imported goods at that time is the discovery by the British archaeologist Michael Hatcher in 1983 CE of ship wreckage loaded with goods from China and Batavia (present-day Indonesia) to Europe that sank in 1640 CE. It contained more than 60,000 pieces of porcelain. Moreover, some historical information about the content of the Portuguese ships (Catarina and Santiago) indicates they were filled with eastern goods not made in Europe. Jorg also found the wreck of the Dutch East India Company called Geldermalsen that contained eastern goods (7, 8).

The records of the East India Company include a letter addressed by the company's board of directors to the person responsible for the East campaign, which states that the engineering designs there far outperformed what was in Europe, hence the keenness to obtain the best of these models. This letter shows that the art of creative design and depth of taste and beauty were higher in India and the East in general compared to Europe (9-11). In other words, the East was superior in its innovative products, designs, refined taste, and content, hence the fascination of the Europeans with such products they could not match. Europeans were fascinated by the technical superiority of the East, as Malachy Postlethwayt pointed out in his voluminous *Universal Dictionary of Trade and Commerce* on the authenticity of the artifacts from the East brought in by the East India Company. The Dutch also imported 43 million pieces of porcelain from the East during the seventeenth and eighteenth centuries, an indication of the global demand for these products unparalleled in Europe (12).

'Mercantilism' as an economic concept emerged to maintain the coinage cycle within the European countries. The concept relied on theories presented by the Italian Antonio Serra in *The Short Treatise on the Wealth and Poverty of Nations* in 1613 CE (13) and refined by the French politician Jean-Baptiste Colbert (1611-1683 CE) and the British economist Thomas Mun 1571-1641 CE, who also developed the concept of trade balance (14-16). However, the term 'mercantilism' appeared for the first time in Adam Smith's *The Wealth of Nations* in 1776 CE (17). The economic concept of 'mercantilism' prompted each European country to put in place several government policies and executive regulations. At the outset, they focused mainly on copycatting and imitation of the products of the East by establishing a local industry and imposing taxes and excise duties on the products from the East to reduce their flow in favor of the imitated or knockoff domestic products. However, the concept expanded two centuries later to reflect the essence of the economic policy of all European countries with each other. There were duties imposed on products coming from France to Britain and vice versa, and from the Netherlands to France and vice versa (18-20). We cannot provide a complete list of these systems here because they multiplied, changed, and were variably applied to different commodities.

This concept led Britain later to get involved in the First Opium War (1839-1842 CE) and the Second Opium War (1856-1860 CE) with China in the nineteenth century. The aim was to force China to buy its products, which Britain imitated (from products imported from the East), but they did not actually have a market inside China. Britain, in order not to lose the trade balance in favor of China, planted opium in the Bengal, brought it into China by force, and sold it in exchange for the Chinese products. When the Chinese emperor banned the opium trade, Britain became enraged and went to war only to allow in opium. Following its victory in the First Opium War, Britain controlled the port of Hong Kong and had reduced taxes on its products and more facilities of their entry to China despite the lack of a market for them due to their lower quality compared to those of the East. Britain obtained even greater concessions after its victory in the Second Opium War, including greater tax breaks and lands and warehouses in the ports. It is worth noting that before the trade relations between Britain and China started, the British King George III addressed the Chinese Emperor (1760-1801 CE) to open trade relations between the two countries. The latter refused because his people did not need the products of the barbarians. It was not a matter of arrogance on part of the Chinese, but the low quality of the European products (21-23).

Europe did not have products to trade with the East. When a Portuguese naval campaign led by Visco de Gama arrived in Zamorin, he asked to meet the King to show him the types of goods available in Portugal that could be the basis for trade exchange between the two kingdoms (the Kingdom of Zamorin of India and the Kingdom of Portugal) on the next trade trips. The King laughed and scorned the poorly made products, while India had better and more valuable products than all the goods offered by the Portuguese. This matter continued for nearly 350 years after the Portuguese invaded India in 1498 CE (24). It recurred later in many situations, including what we mentioned above about the British desire to trade with China. Accordingly, during that time, Europe had nothing but to adopt the policy of imitation of the products of the East, imposing taxes, and introducing its inferior products to the markets of the East, even if through war, along with privileges, tax exemptions, control of ports and warehouses to make its goods available to the eastern consumer. The Europeans went as far as spreading drugs to maintain the trade balance with other countries. Although imitation did not make European products equal the quality, beauty, and taste of the products of the East, it reduced the outflux of silver currency to the East and generated a local consumption of its products, especially for the ordinary European consumer because Eastern products were still in high demand among the ruling class and the aristocracy (1).

In short, all the political, economic, and social conditions in the European countries encouraged imitation of the products of the East. Each imitation was dubbed innovation, and imitators (seen as innovators in Europe) were supported by the rulers first and then by the integrated systems and institutions created to support and legitimize imitation. Later in this book, we will elaborate on the patenting systems in their two stages (referred to at the outset of this chapter). Both stages were not related to innovation or encouraging local innovators. Instead, they were a purely economic process to support and encourage imitation of other nations' innovations, as we will mention later.

# References

1. M. Berg, In Pursuit of Luxury: Global History and British Consumer Goods in the Eighteenth Century. *Past & Present*, 85-142 (2004).

2. M. Berg, From Imitation to Invention: Creating Commodities in Eighteenth-Century Britain. *The Economic History Review* 55, 1-30 (2002).

3. B. Godin, Innovation: the history of a category Working Paper. *Institut national de la recherche scientifique, Centre Urbanisation Culture Société, Montréal* (2008).

4. A. Gerard, *An Essay on Taste (1759): Together with Observations Concerning the Imitative Nature of Poetry*. (Scholars' Facsimiles & Reprints, 1978).

5. W. Hogarth, *The Analysis of Beauty: Written with a View of Fixing the Fluctuating Ideas of Taste*. (W. Strahan, 1772).

6. G. Tarde, *The Laws of Imitation*. (Read Books Limited, 2013).

7. G. L. Miller, The Second Destruction of the Geldermalsen. *Historical Archaeology* **26**, 124-131 (1992).

8. C. Sheaf, R. Kilburn, R. S. Kilburn, *The Hatcher Porcelain Cargoes: The Complete Record*. (Phaidon, 1988).

9. E. I. Company, F. C. Danvers, *Letters Received by the East India Company from Its Servants in the East*. (S. Low, 1900).

10. K. N. Chaudhuri, *The English East India Company: The Study of an Early Joint-stock Company 1600-1640*. (Routledge, 1999).

11. P. Lawson, *The East India Company: A History*. (Taylor & Francis, 2014).

12. M. Postlethwayt, *The Universal Dictionary of Trade and Commerce, with Large Additions and Improvements, Adapting the Same to the Present State of British Affairs in America Since the Last Treaty of Peace Made in the Year 1763; with Great Variety of New Remarks and Illustrations*

*Incorporated Throughout the Whole: Together with Every Thing Essential that is Contained in Savary's Dictionary: Also, All the Material Laws of Trade and Navigation Relating to These Kingdoms and the Customs and Usages to which All Traders are Subject.* (W. Strahan, 1774).

13. A. Serra, S. A. Reinert, *A 'Short Treatise' on the Wealth and Poverty of Nations (1613).* (Anthem Press, 2011).

14. T. Mun, *England's Treasure by Foreign Trade: Or, the Balance of Our Foreign Trade is the Rule of Our Treasure. Written by Thomas Mun.* (R. and A. Foulis, 1755).

15. T. Mun, G. J. Adams, N. Page, *Thomas Mun: The Complete Works (Economics and Trade).* (Newton Page, 2013).

16. J. B. Colbert, *Memoirs of the Marquis of Torcy, Secretary of State to Lewis XIV. Containing the History of the Negotiations from the Treaty of Ryswic to the Peace of Utrecht. Translated from the French.... of 2; Volume 2.* (Creative Media Partners, LLC, 2018).

17. A. Smith, *An Inquiry Into the Nature and Causes of the Wealth of Nations.* (Legrand, 1791).

18. M. Lange, J. Mahoney, M. Vom Hau, Colonialism and development: a comparative analysis of Spanish and British colonies. *American Journal of Sociology* **111**, 1412-1462 (2006).

19. D. Ormrod, O. David, *The rise of commercial empires: England and the Netherlands in the age of mercantilism, 1650-1770.* (Cambridge University Press, 2003), vol. 10.

20. I. Wallerstein, *The modern world-system II: Mercantilism and the consolidation of the European world-economy, 1600–1750.* (Univ of California Press, 2011), vol. 2.

21. A. Waley, *The opium war through Chinese eyes.* (Routledge, 2013), vol. 33.

22. P.W. Fay, *The Opium War, 1840-1842: Barbarians in the Celestial Empire in the Early Part of the Nineteenth Century and the War by which They Forced Her Gates Ajar*. (University of North Carolina Press, 2000).

23. J. Lovell, *The Opium War: Drugs, Dreams and the Making of China*. (Picador, 2012).

24. R. Crowley, *Conquerors: How Portugal seized the Indian Ocean and forged the First Global Empire*. (Faber & Faber, 2015).

# CHAPTER TEN

CHAPTER TEN

# REPUBLIC OF VENICE: EUROPE'S MODEL FOR GRANTING "LETTERS OF PATENTS" CONCESSIONS

As the concept of innovation relates to patents, this part of the book covers patents from the fourteenth to the nineteenth century. We will elaborate on how patents related to imitation and copycatting rather than innovation as is common today. The start will be with the Italian republics of Venice and Florence (in this chapter) and then with the Netherlands, Britain, and France (in later chapters).

In 1416 CE, Sir Francis Petri (a foreigner from the Greek island of Rhodes in the Mediterranean) was granted a patent for rubbing wool. However, this method was not an innovation but rather in place in Rhodes, whose residents were famous for quality wool processing. Accordingly, the Venetians sought to bring this artisan to Venice to transfer the method, which was not an innovation but new and unknown to the Venetians. It was also more helpful industrially than Venice's method in the wool industry. Through this document (patent), Petri obtained the concession to manufacture wool by the method in Venice (1, 2).

Likewise, Filippo Brunelleschi (the engineer who designed Florence Cathedral) was granted a three-year patent concession in Florence in 1421 CE for a river transport vessel specially made for loading marble. Although it sank shortly before completing its first voyage, such river transport vessels were mentioned in the writings of the ancient Greeks and Romans (1, 3-5).

In 1449 CE, Venice again granted a 20-year patent concession to John of Utynam that granted him a monopoly on the making of stained glass, which was brought to Venice from Egypt and the countries of the East. However, the transparent stained glass industry did not exist in Venice; it was new as an industry in Venice but not in the world. Accordingly, modernity and innovation had a narrow perspective for those who did not know the innovations around them; patents were granted for imitating existing innovations (6-8).

The Republic of Venice granted 13 patents from 1416 to 1472 CE (before issuing the patenting system), three of which were for one person (of French origin) called Antonio Marini. All these patents, though in different fields, dealt with only one economic issue: finding ways to knock off and imitate the products of the East. These were not substitutes for imports but rather an attempt to reproduce the exact imported product, such as, porcelain, glass, silk, and wool, or their industrial methods and processes (9-11).

The concept of patent concession spread from Venice to the rest of Europe, which depended on imports from Britain, France, and the Netherlands in particular. In 1555 CE, King Henry II of France (1547-1559 CE) granted Abel Foullon a patent for his holometer, an instrument for making angular measurements for surveying and construction, thus monopolizing its manufacture and sale for ten years within France. This instrument was foreign to the French but existed in the East (6, 12-14).

Again in 1557 CE, King Henry II granted a design and printing patent concession to a designer named Robert Granjon though Venice granted this same patent in 1502 CE to Aldus Manutius. In other words, the same method was considered an innovation in France 55 years after the patent granted by Venice to its designer. Most of these designs were present in the East. Venice granted a patent to a designer who was well-versed in imitating the designs of the East. Then the French granted another designer a patent for his ability to imitate the designs of the Venetians. This process was inherent in the innovation chain (15-18).

In the same period, Britain tried to imitate the concept of Venice, as did France and the Netherlands. Edward VI of England (1547-1553 CE) granted a patent concession in 1552 CE to Henry Smyth to manufacture the stained glass he copied from Venice (19-21). Queen Elizabeth I (1558-1603 CE) was very active in this field. She granted patent concessions to fifty people from 1561 to 1590 CE based on her personal opinion and feedback from the royal court about the benefits of the innovation to the country's economy. Britain was suffering from hyperinflation; the price of goods rose 6.8 times from 1510 to 1597 CE, due to the Spanish silver brought from the American colonies. The decline of the Crown's revenues to 40% forced the Queen to increase borrowing to cover expenses. The royal court saw that the concession rights would generate returns to the Queen and a fixed revenue from the annual fees paid by those who obtain the concession paid to the British Crown directly at the grant of the patent or on an annual basis as per the agreement. The pre-union Dutch provinces and the disintegrating Roman Empire states also granted several patents (11, 20).

In 1945 CE, Flunmkin Zoo pointed out that what was referred to as patent concession rights since the era of Edward III (1327-1377 CE) was an untrue legend. The claim that John Kempe was granted a patent concession in Britain in 1331 CE has not been proven (5). It is good we have some vigilant historians who expose the attempts to misrepresent history and distort events for the benefit of a nation, a state, or a king. This book supports what Flonmkin said that the alleged 1331 CE patent was unreal. Logically speaking, how did the practice of patenting stop for more than a hundred years on the one hand? How was the patent introduced as a precedent that Europe did not know and was not induced by the civilizational shocks presented by the Venetians through the Silk Road, nor what the Spaniards and the Portuguese brought through the new maritime trade routes?

The prevalent economic concept led by Venice was to encourage imitation and copycatting of the products of the East and their industrial methods to protect the economy and maintain coinage circulation. We cannot list here all the intellectual property concessions granted by the kings and dukes of Europe during this period due to their abundance. Additional examples will further show the banality and naivety of some concessions granted.

Take, for example, the salt concession in England. In the sixteenth century, England depended on a few salt marshes and imported salt from the Guyenne region in southern France on the border of Spain with a coastal outlet to the ocean. When France and England were at war, they agreed in 1530 CE that the French and the heirs to the French throne would provide England with 140 thousand bushels[5] of salt for forty years. When disputes reignited between the two parties, salt became scarce in England, and it was impossible to import it from Guyenne (22).

Under these circumstances, Queen Elizabeth I granted a patent in 1564 CE to the German Jasper Seler to establish a salt factory in Britain. He had a salt monopoly for 20 years, provided that it is not sold to a foreign entity. He had to pay 10% of the profits to the Queen and sell the white bushel for 8 pence[6]. Granting the patent had nothing to do with innovation but just a way to attract a foreign investor to invest in England and transfer knowledge there. This knowledge was not even an innovation. It was a process for heating and filtering seawater.

5 A bushel was a unit of weight equal to 30.027 kilograms. It was commonly used to weigh foodstuffs, such as wheat, salt, etc.

6 A penny was 12 shillings, and the sterling was 20 shillings. That is, one pound sterling is equal to 240 pence.

However, granting the concession was to bring investors' capital to England to reinvigorate the English economy (22).

British history is full of examples of patents granted with nothing to do with innovation but rather a mechanism to imitate imported products. In 1571 CE, Richard Mathew obtained a patent and the right to manufacture knives, the same as Ottoman knives. He even carved the crescent on the knife to prove it was a good imitation of the Turkish quality. In a later chapter, we will tell the story of creating an entire city for weapons and knives in Britain to imitate and copy industries abroad. Versaline obtained a patent in 1574 CE to monopolize the imitation of glass imported from Venice (23).

The intellectual property concessions from the Duke in Venice were not a precedent. History includes various similar cases. The Spaniard Benjamin of Tudela, a traveler between East and West (1160-1173 CE), mentioned that he witnessed the King of Palestine granting an exclusive concession to whoever would come up with a method for imitating the production of textile dyes, as they were precious and expensive while their extraction and mixing process involved profound craftsmanship. Perhaps the Venetians borrowed this process from Palestine (5).

If we go back further in history, the historian Phylarchus of Naucratis in the third century BC mentioned in his book

*The Learned Banqueters* that in the sixth century BC, there was a city called Sybaris in southern Italy under the control of the Greeks on land and sea trade route at that time. It was thriving because its chefs were granted a one-year concession for any new dish added to the menu. This example may be the oldest precedent recorded for preserving intellectual property. Others may have preceded it, but they have not yet reached us or discovered (14, 24, 25).

# References


1. P. O. Long, Invention, secrecy, and theft: Meaning and context in the study of late medieval technical transmission. *History and Technology* **16**, 223-241 (2000).

2. D. L. Bohan, Computer Programs: Abstract Ideas or Patentable Subject Matter. *Suffolk UL Rev.* **29**, 809 (1995).

3. F. D. Prager, Brunelleschi's patent. *J. Pat. Off. Soc'y* **28**, 109 (1946).

4. I. Mgbeoji, Patents and Traditional Knowledge of the Uses of Plants: Is a Communal Patent Regime Part of the Solution to the Scourge of Bio Piracy? *Indiana Journal of Global Legal Studies* **9**, 163-186 (2001).

5. M. Frumkin, The orgin of patents. *J. Pat. Off. Soc'y* **27**, 143 (1945).

6. R. A. Klitzke, Historical Background of the English Patent Law. *J. Pat. Off. Soc'y* **41**, 615 (1959).

7. M. Luis Ferreira Nascimento, The First Patents and the Rise of Glass Technology. *Recent Innovations in Chemical Engineering (Formerly Recent Patents on Chemical Engineering)* **9**, 20-30 (2016).

8. A. Gomme, Early British Patent Grants. *Nature* **119**, 494-494 (1927).

9. G. Mandich, Venetian Origins of Inventors' Rights. *J. Pat. Off. Soc'y* **42**, 378 (1960).

10. G. Mandich, Venetian Patents (1450-1550). *J. Pat. Off. Soc'y* **30**, 166 (1948).

11. C. A. Nard, A. P. Morriss, Constitutionalizing Patents: From Venice to Philadelphia. *Review of Law & Economics* **2**, 223-321 (2006).

12. M. Franzosi, in *Intellectual Property in Common Law and Civil Law.* (Edward Elgar Publishing, 2013).

13. M. E. Valentinuzzi, Patents and scientific papers: Quite different concepts: The reward is found in giving, not in keeping [retrospectroscope]. *IEEE Pulse* **8**, 49-53 (2017).

CHAPTER ELEVEN

# THE REPUBLIC OF VENICE: THE FIRST INSTITUTIONAL PATENTING SYSTEM

The excessive patent concessions granted in Europe as an individual procedure based on the ruler, the king, or the duke's sole opinion created an imbalance. These concessions were misused, especially in Britain, during the reign of Queen Elizabeth I and her successor, James I (1603 1625 CE) (1-3). Merchants and citizens in Europe started to complain.

The Republic of Venice, which took the lead in granting individual concession rights in Europe, or the so-called "letters of patents," was the first European country to turn to an institutional patenting system not based on the duke, ruler, or king's opinion. Venice issued its patenting system in 1474 CE. The demand for such a system arose out of the struggle of interest groups from the public, traders, and investors rather than the awareness of legislators and political regimes. So, the patenting system came as an expected result of solving the crisis of improvised individual power abuse not based on rules and regulations (4-7). We need to understand the economic and social situation of the Republic of Venice, which prompted the issuance of the patenting system.

The economic system in Venice relied on the law of 1173 CE, which codified the relationship between the government and the guilds of artisans and merchants. Even with the import of specific goods, the trade practice in Venice was a profession that had rights. It was forbidden for foreigners or citizens outside the profession to practice it without the state's permission. The population of Venice ranged from 100,000 to 150,000, spoke Greek, and welcomed foreigners. The income tax was low (5%) compared to the rest of Europe. Citizens comprised three social classes: the patrician (2%), the cittadini (5%), and the commoners (93%) (4, 7, 8).

Patents allowed citizens or foreigners outside the guild to engage in commerce. The patenting system issued in 1474 CE curtailed the powers of the guilds under the law of 1173 CE. It also aimed to diversify the industry and attract foreigners to transfer their expertise and skills to Venice. The patent did not allow monopolizing a commodity, industry, or industrial process. Instead, it allowed its holder to practice a commercial activity within the relevant guild (4, 7, 8). Complaints by foreigners and Venetians abounded due to the reproduction of their goods or industrial methods (borrowed from abroad or brought with them from their countries of origin) by the rest of the Venetian artisans and merchants. Eventually, this led to the issuance of the Venetian Patent Statute of 1474 CE. As per this system, the Grand Council granted the authority to an entity called the Prorreditoridi Commun to grant patent rights and monopoly protection for ten years for everything innovative, original, and useful (4).

An in-depth analysis of this system would show that the terms' innovative', 'original,' and 'useful' had different meanings from today's definitions in modern patent systems. 'Innovative' meant new to the residents of Venice, even if the product was invented hundreds of years ago and imported and traded in the Venetian markets. It was only necessary to prove that it was not made in Venice (whether it was an innovative product or industrial process). Whoever brings it to Venice or learns it abroad and returns to apply it in Venice was granted a 10-year exclusive monopoly right for manufacturing, production, and sale. The issued patent stated that right under the laws of the Republic.

In *Owning Ideas: The Intellectual Origins of American Intellectual Property*, Oren Bracha explained the legal differences between modern and old innovation systems. Although his exposition relates to only the US and British systems, it helps understand the differences in past and present legal definitions and expressions concerning patents (9).

The staff of the relevant guild examined prior art for innovation to answer the following questions: is the technique/method available in Venice today? Is it useful in and applicable to the industry? Accordingly, the patent was granted (4, 7, 10).

These innovations were imitations of products imported and sold in Venice (or carried by Venetian merchants on the Silk Road) but not made in Venice. It also applied to industrial methods unavailable in Venice but found in other countries. This explains the passion of the Venetian merchants to travel from the thirteenth to the eighteenth centuries to learn about new industries and innovations in the East. Marco Polo wrote a detail of his travels in which he tried to learn about the new nations in the East. He also reported the story of his father's attempt to steal an industry secret from China and transfer it to Venice (11).

Accordingly, the patent system was a 'pure' imitation system not based on innovation. Still, it was an efficient and practical institutional system that led to the transfer of many industries and innovations from the ends of the world to the Republic of

Venice, including glass, silk, porcelain. Venice became a model for European countries in the following four centuries.

This turned Venice in the fifteenth and sixteenth centuries into the most prominent industrial gathering in Europe. Venetian and foreign entrepreneurs brought their ideas and distinctive crafts to the markets and factories of Venice, which provided access to markets between East and West through a developed trading network of Mediterranean seaports. Imitated ideas flourished in Venice faster than anywhere else in Europe (12, 13).

Pamela Long indicated that the Venetian patenting system was created and developed to control crafts, knowledge, and skills. It was a reward to foreigners for transferring their skills to a country or a reward to citizens for introducing new methods. Long believes that the patent system revolved around knowledge related to crafts, such as glassmaking, pottery, and others. This system encouraged the transfer of artisanal knowledge and its mechanisms to a new country (14).

Despite our admiration and support for Long's proposal that the patent system was mainly to serve artisans and their guilds, which many historians of technology and patent laws missed, his proposal did not encompass the innovation concept used in that era to denote imitation. This book addresses this issue.

Maximilian Frumkin pointed out that there were no patents for inventions in ancient times. He provided an in-depth analysis in criticism of the so-called patents in European countries, stating that most patents were not for innovations (15).

To better analyze the patents in Venice since the issuance of the patent system in 1474 CE to the end of the Republic in 1797 CE, there were two systems or mechanisms for granting patents: fixed and customized. The fixed system was faster and under the authority of an administrative body called the Prorreditoridi Commun. It granted 10-year patents when specific criteria for innovation were met (as indicated above). However, the customized patent was granted by the Senate on a case-by-case basis from ten up to sixty years. Some patents would last throughout the inventor's life or even his heirs (extended protection for two generations). From 1560 to 1580 CE, the average patent period was 25 years, and patents in energy and water usually had a protection of 5-7 years more than other fields. The innovation infringement fine was set in the patent and ranged from 2 to 1,000 ducts[7]. The violator was also forced to compensate for the losses and remove the manufactured products from the market (4, 16).

7 Duct was a coin made of gold (99.47% pure) for the inhabitants of Venice issued in 1284 CE and weighed 3.545 grams. Before this date, the Venetians used the duct currency of the Byzantine Empire in their commercial dealings.

Let us take samples of specific patents to know the difference in the duration of grants, fines, and conditions. Sebastiano di Silvestri was granted a 50-year patent and a fine of 50 ducts for infringement. Other innovations were granted 25-year protection with conditions such as not impeding streams and rivers, which were imposed in one of the patents as a condition for granting (16).

Surprisingly, the patent system in Venice outperformed the current patent systems regarding customized patents as per economic importance and benefit to society. Contemporary innovation systems have the same advantages and penalties for all industry categories though each category has its specialization and different considerations when innovating, such as effort, time, cost, and many other factors. The pharmaceutical and computer software industries are current examples in this regard (16).

In their 2009 book, *Patent Crisis and How the Courts Can Solve It*, Burk and Lemley pointed out that the US patent system involves significant shortcomings and a dilemma due to the standardization of procedures despite the different nature of patent applications. They suggested customizing different protection systems to suit each case and industry (17). In 2019, Stefania Fusco pointed out the lessons the world and the USA could learn from the Venetian customized patents more than

CHAPTER TWELVE

# DIFFUSION OF PATENTING SYSTEMS IN EUROPE: THE NETHERLANDS, ENGLAND, AND FRANCE

The second patent system in Europe was issued in the Netherlands in 1580 CE after the seven independent United Provinces of the Netherlands, commonly referred to as the Dutch Republic in historiography, took shape. The Dutch Golden Age extended from 1581 to 1672 CE. Before 1580 CE, every independent province issued concessions for innovations (imitations, so to speak). The patent system was for the Netherlands, while each province and some large cities had their systems. There were three main types of patenting in the Netherlands in the sixteenth and seventeenth centuries with different criteria, conditions, and examination procedures (1, 2).

The analysis of the various patent systems in the Netherlands requires a review of the preceding economic and social conditions, which paved the way. The Netherlands advanced economically first because of sea trade with the East. It is reported that any visitor to the Netherlands in the sixteenth and seventeenth centuries would find all the products of the East and West in the Dutch markets, which caused a remarkable economic boom; even some Dutch cities, such as Utrecht and Amsterdam, became the center of admiration of Europeans. The Netherland cities outperformed the Italian cities that were the source of oriental trade and technologies imitated and reproduced in Europe. An English traveler reported that "*the Netherlands has surpassed the seven wonders of the world.*" He was impressed by the quality of architecture and urbanization and mentioned a forest of underground water and sewage pipes, as they were superior to their counterparts in other European cities. Outside Europe, the Dutch engineers and artisans were in high demand. Kings, such as the Russian Tsar Peter, sent a group of Russian engineers study in the Netherlands (2-4).

The old patent records indicate that the average number of annual patents was seven from 1590-1640 CE for different protection periods ranging from five to twelve years. Concerning the mechanism for examining innovations, originality or novelty was not a condition for granting a patent. All imported products were patented to localize industries, as the Venetians did for nearly a hundred years before, and as the English and French would do for two hundred years after. Analysis of the Dutch patenting (or imitation) model requires a review of the most active categories in obtaining patent monopoly concessions (1).

Each craft (carpenters, masons, etc.) in the Dutch cities had a chief, a government employee supervising public works. They obtained more than one-fifth of the total patents. Some of them were appointed as chiefs in their respective cities because they were inventors and to apply their ideas. Innovation was not limited to the aristocracy or the wealthy due to the high costs of patent fees. Artisans were among the most active in filing patent in the Netherlands. One-third of the patents granted in 1580-1720 CE was for workers and wood and metal artisans. An examination committee assessed each invention and used external experts if necessary to know the validity or applicability of ideas (1, 5).

The specification of innovation would be a written description, drawings, or a working model. Likewise, a third of the patents from 1600-1640 CE was for hydraulic devices (because the Dutch cities were amid a network of rivers and canals), industrial mills, and thermal heating devices; 1600-1640 CE witnessed the peak of patenting in the Netherlands then declined not due to the economy's deterioration and the unwillingness to imitate foreign products, but because the patent system developed and became more sophisticated in terms of examination and granting. One patent required the applicant to place a model in the river for one year to prove his idea's feasibility and that it was better than the existing ones (1).

In England, after the period in which the king or queen granted patents as per their personal opinion about the usefulness of the idea brought from abroad ended, King James I faced a challenge and pressure from the people and the Parliament until he agreed in 1624 CE to issue the first patent system in Britain. The King continued to issue patents, other than those issued by the Parliament, as per the new system. In other words, there were two channels for patenting in England until King James I was forced to stop this practice. The patenting manipulation combined with other economic causes (referred to partially above) triggered the English revolution in 1642-1651 CE, which led to the execution of King Charles I (1625-1649 CE) on charges of treason at the hands of the rebel leader Oliver Cromwell, who ruled the country until he died in 1661 CE. After his death, the Parliament called the heir of the executed king, King Charles II (1660-1685 CE), from exile to rule under the supervision of Parliament. Only one channel for patent issuance remained after that (6-8).

We do not claim that issuing patent concessions by the king was the only cause of the revolution, but it was significant cause. The arbitrary patenting led to the monopoly of the import and manufacture of goods, some of which were present in Britain and available, though in small quantities, in its markets, hence the destabilization of the economic situation ensued by general unrest of the people over the high prices and of merchants over the stagnation of their trade or preventing them from expanding for the benefit of only two parties, namely the patent holders granted the concessions for manufacture or importation, and secondly, the royal court, which received fees for granting and maintaining the concessions, a steady and growing source of income for the ruling family (9, 10).

The 1624 CE decision restricted the royal privileges in granting patent monopoly rights. It marked the start of institutional intellectual property in Britain, though the royal court did not abide by these laws until the outbreak of the English Revolution. The Parliament's decision was a reaction to the abuse of the royal powers. The patent system in Venice was more advanced in granting patents, more precise in implementing protection, and less arbitrary.

Analyzing the economic pros and cons of patents in England before the patent system requires us to understand the rationale of the Letters of Patents granted by the royal court and the public's reaction. Even though patents issued directly by the king improved the economy in some areas and periods, the increasing monopoly rights eventually infringed on merchants' rights in the English markets. They had a clear adverse impact on the economy. Many merchants complained about the harm they sustained due to the patents, thus prompting the English Parliament to address the monarchy and restrict their powers during the reign of James I. However, the problem began at the end of the reign of Queen Elizabeth I. Granting the royal letters of patents was to control the conduct of trade by giving monopoly rights, a response to the government's inability to finance its projects. Thus, the royal letters of patents were: 1) to monopolize industry and 2) to monopolize importation franchise (10).

The monopoly was to induce the import of new technologies new to Britain though known for hundreds or thousands of years outside it. This also applied to the import of new goods that did not exist in the domestic markets. The damage began when the royal letters of patents included industries that actually existed in Britain, or the import of goods that were actually available in the British market, even in small quantities. Patents were also issued by the royal court for the benefit of foreign investors or artisans to attract their money or skills. The increasing number of patents harmed the national artisans and investors. Those affected by monopoly rights complained to Queen Elizabeth at the end of her reign (because patents were issued directly by the Queen and no one had the right to object to her decision), she allowed grievance cases to be decided in the general courts, provided that the royal court was consulted in the ruling (9, 10). We should not forget that stopping or suspending a patent granted (by a court ruling, for example) had a negative impact on the income of the royal court, as this practice was a major source of revenue.

Regarding France, as we noted at the outset of this chapter, King Henry II granted patents to engineer Abel Foullon, appointed after Leonardo da Vinci (11, 12). The French monarchy continued to grant letters of patents for about 250 years. It was also a significant source of revenue for the French royal court, which fiercely defended this right until the French Revolution in 1789 CE. King Louis XVI (1774-1792 CE) signed the French patent system forcibly in 1790 CE, two years before the French revolutionaries executed him. From the issuance of patent letters by the king to the patent system, this period witnessed various phases and changes summarized here (13, 14).

In 1670 CE, Minister Jean Baptiste Colbert criticized the mechanism of issuing and granting letters of patents during the reign of King Louis XVI. He pointed out it lacked clarity, specifically in textile and porcelain technologies. The inventor was anyone who would create an industry inside the country that was similar to or imitative of an industry outside the country or imported into French territory (15, 16).

For example, the King granted a patent to Mr. Milne to transfer the English wool weaving industry to France, with a bonus of 60,000 livres in advance and 6,000 livres annually. The Irishman Holker was granted a patent concession to build an English wool factory with a reward of 90,000 livres. Wool was imported from England, and the King desired to grant concessions with incentive bonuses to motivate foreign artisans to come to France and set up their factories there. The concession had nothing to do with innovation in France either. As per Boufflers' description, a member of the French Academy of Sciences, 90% of the royal letters of patents in France were for foreigners to transfer their industries to France under the name of patents (15, 17, 18).

France's first industrial property rights system was issued in 1712 CE (modified and complemented up to 1744 CE). It was an adaption of the Venetian system issued in 1173 CE, but with devastating modifications on the French economy. This law stipulated that a "*royal concession*" must be granted by artisans to produce any goods in French factories. It included every industrial method, such as weaving or spinning, was a property that must be protected by law, and only the king could grant this protection! Accordingly, a fee must be paid to the royal court when selling in the market. This system overlapped with the concepts of the intellectual property until a decision was issued for the inventor's right of ownership in 1762 CE. Before that, the inventors did not have intellectual property as per the applicable system. The king only compensated the innovator (or the imitator here) with a reward the king would determine. Then the industrial property rights system was abolished in 1774 CE, thus paving the way for the French Revolution. After the revolution, the French patent system was issued in eighteen chapters, which are helpful to understand the concept of imitation and innovation (15, 19).

After the issuance of the new patent system proposed by Boufflers and approved in 1790 CE, imitation and copycatting were still at the core of the system, despite the resonant literary phrases used to describe the innovator's natural right. What about the natural right of the imitator to imitate him? Patents were governed by the French Academy of Sciences and granted for 5, 10, or 15 years, depending on the best interest as decided by the Academy. The patent would be revoked if not used within two years from the grant date or if registered in any foreign country (20, 21).

Also, the new French patent system (1790 CE) recognized a type of import patent, which was actually in effect 300 years ago in Venice and 150 years in England. Import patent was an exclusive concession granted to import a foreign commodity into French territory and was similar to the franchise system in force in some developing countries today. In France, the system guaranteed the issuance of a patent for anyone who brings a foreign innovation to France or imitates it if he was the first to introduce it to French territory and if it was not present in the French market at the time of the patent application (22).

This system was a legalization of what was previously in place in France and Britain. The import-related patent was the most problematic and the subject of severe criticism. Some patent holders disputed the merchants who had previously imported the same commodity and brought it to the French and British markets. It was a cause of economic instability that eventually led to the revolution against the monarchies in both countries because the peoples and parliaments accused the royal court of colluding with foreign investors and merchants and receiving fees for patents granted to benefactors. Even within both countries, the king's privileged circle or the investors and aristocracy close to him could pressure the royal court to obtain a patent for a monopoly on the import of a commodity or an industry. This serious practice eventually led to the fall of the two monarchies in both countries. The monarchy in Britain did not return until after the Parliament seized the right to decide the fate of the country, while in France the monarchy never returned after the revolution.

# References


1. K. Davids, Patents and Patentees in the Dutch Republic between c. 1580 and 1720. *History and Technology, an International Journal* **16**, 263-283 (2000).
2. D. S. Landes, J. Mokyr, W. J. Baumol, *The Invention of Enterprise: Entrepreneurship from Ancient Mesopotamia to Modern Times. Chapter 6: The Golden Age of The Dutch Republic.* (Princeton University Press, 2012).
3. M. North, C. Hill, *Art and commerce in the Dutch Golden Age.* (Yale University Press New Haven and London, 1997).
4. M. Prak, Guilds and the Development of the Art Market during the Dutch Golden Age. *Simiolus: Netherlands Quarterly for the History of Art* **30**, 236-251 (2003).
5. C. Florea, From Venice to the Dutch Republic. *Optics and Photonics News* **29**, 42-49 (2018).
6. T. K. McBride Jr, Patent Practice in London-Local Internationalism: How Patent Law Magnifies the Relationship of the United Kingdom with Europe, the United States, and the Rest of the World. *Loy. U. Chi. Int'l L. Rev.* **2**, 31 (2004).
7. C. MacLeod, *Inventing the Industrial Revolution: The English Patent System, 1660-1800.* (Cambridge University Press, 2002).
8. T. T. Ochoa, M. Rose, The anti-monopoly origins of the patent and copyright clause. *J. Copyright Soc'y USA* **49**, 675 (2001).
9. A. Mossoff, Rethinking the development of patents: An intellectual history, 1550-1800. *Hastings Lj* **52**, 1255 (2000).
10. C. A. Nard, A. P. Morriss, Constitutionalizing Patents: From Venice to Philadelphia. *Review of Law & Economics* **2**, 223-321 (2006).
11. M. Frumkin, The orgin of patents. *J. Pat. Off. Soc'y* **27**, 143 (1945).

12. R. A. Klitzke, Historical Background of the English Patent Law. *J. Pat. Off. Soc'y* **41**, 615 (1959).

13. F. D. Prager, A History of Intellectual Property from 1545 to 1787. *J. Pat. Off. Soc'y* **26**, 711 (1944).

14. J. Baudry, "A Politics of Intellectual Property: The French Revolution and the Creation of a Patent System,» (2020).

15. F. Neumeyer, Contribution to the history of modern patent legislation in the United States and in France. *Scandinavian Economic History Review* **4**, 126-150 (1956).

16. E. C. Spary, *Eating the Enlightenment: Food and the Sciences in Paris, 1670-1760.* (University of Chicago Press, 2013).

17. J. R. Harris, *Industrial Espionage and Technology Transfer: Britain and France in the 18th Century.* (Taylor & Francis, 2017).

18. G. Galvez-Behar, The Patent System during the French Industrial Revolution: Institutional Change and Economic Effects. *Jahrbuch für Wirtschaftsgeschichte / Economic History Yearbook* **60**, 31-56 (2019).

19. L. Hilaire-Pérez, Technology as a public culture in the eighteenth century: The artisans' legacy. *History of science* **45**, 135-153 (2007).

20. M. Crosland, A. Galvez, The emergence of research grants within the prize system of the french academy of sciences, 1795-1914. *Social Studies of Science* **19**, 71-100 (1989).

21. F. D. Prager, Proposals for the Patent Act of 1790. *J. Pat. Off. Soc'y* **36**, 157 (1954).

22. E. Burke, Reflections on the Revolution in France (1790). *Everyman's Library*, (1986).

PART FOUR

# USA: THE COUNTRY THAT ROSE ON IMITATION AND COPYCATING

# CHAPTER THIRTEEN

## CHAPTER THIRTEEN

# REPLICATING AND IMITATING INNOVATIONS IN THE AMERICAN COLONIES BEFORE THE INDEPENDENCE

Copycatting and imitation as an approach and policy in Europe reflected in its American colonies, where the imitation process in their mother European countries continued. The American colonies followed the example of the countries of origin concerning copycatting and imitation but with a higher dose, especially after the independence.

The economic policy in Europe was that raw products were brought from the American colonies, manufactured in Europe, and then resold to the colonies. When the USA was established, it suffered from a scarcity of advanced industries. Some independence advocates warned that the USA, after its independence, would not withstand for long because it was dependent on Europe in production and industry. Thus, following the independence, the policies and regulations tended toward imitation, copycatting, and replicating what existed in Europe (1).

This chapter will begin with a tale showing how some innovations (or imitations in the USA) originated. Then, we will analyze the patent-related systems for analyzing innovation and imitation in the American colonies before the independence from the English colonization of the eastern coasts of North America from the seventeenth century to the Declaration of Independence in 1776. Then, (in the next chapter), we will learn about copycatting and imitation since the Declaration of Independence, the opinion of the political leaders encouraging such practice, and review the relevant regulations and patents. Then we continue (in the chapter that follows it) to present models of imitation and replication that accompanied the US progress in World War II and then in the Cold War up to contemporary history. In short, the USA is the largest imitative country in the world today. The author does not believe this description applies to China, as this chapter will show.

In 1621, King James I (1603-1625), King of England and Scotland, needed weapons and swords for the Thirty Years' War (1618-1648). After he used to buy weapons from Prussia and Germany, he brought 12,000 artisans to carry out this task. Accordingly, he granted a patent in 1621 to Thomas Murray, the maker of knives for the Prince of Wales. During the reign of Charles I (1625-1649), precisely in 1629, the King invited many German swordmakers to immigrate to England, built the city of Hounslow for them, and provided them with patents to monopolize this industry (2, 3).

There was a blacksmith and a swordsmith named Benjamin Stone at that time. He assembled knives for the London Cutlers Company, which bought the sharp blade from Italian and German merchants. In July 1636, Stone obtained a 14-year patent from King Charles I to manufacture swords and blades against 20 shillings per year in return paid to the royal court. He resided in Hounslow, the manufacturing hub of knives and swords located on a river canal. The blacksmiths and swordmakers built factories (or mills) on the canal's edge to leverage the water flow to move the mills that sharpened the metal tools. German artisans and swordsmiths built their workshops or factories at the edge of canals and rivers to benefit from water movement. These were not innovations in the manufacture of mills; this practice spread on the banks of German rivers before it spread to Hounslow. Benjamin Stone's letter to the

King did not claim any innovation but rather indicated that he had learned and mastered the manufacture of sharp blades and could produce them in high quantities and quality comparable to the Germans' (3).

Joseph Jenks (1599-1683) worked for Benjamin Stone, who obtained the patent above in 1636 to manufacture swords for 14 years in England, Ireland, and Wales. The patent also included the repair of Ottoman' musket' rifles. This patent was not for innovation, as we have indicated, but a desire from the King to localize the German swordmaking technique in the English city of Hounslow (4).

After mastering Benjamin Stone's German swordmaking, Joseph Jenkins left Britain to the new American colonies, wishing to procure a patent similar to his former employer's in Britain. He benefited from the fact that this technology was unknown in the new colonies. He obtained the monopoly right in Massachusetts, one of the new colonies for a process that resembled that of his mentor, Stone. It was just the transfer of an existing item to a new land that did not know it. He obtained several patents for products, some of which were present and sold in European markets and reached the markets of the new colonies, including a plowing tool used thousands of years ago by the peoples of Mesopotamia and ancient Egypt. One of Joseph Jenkins' innovations included a windmill he built on the river when he came to the new colony to take advantage of the river's flow to operate factory or workshop mills (4, 5). The construction of such workshops was widespread in Europe, and it was neither modern technology nor an innovation.

Joseph Jenkins learned from Stone and transferred the craft to the American colonies; there was no 'innovation'! The patent right was granted to encourage foreigners to transfer non-existent industries by giving them the right to manufacture for a limited time. This issue recurred a lot after and before this date in the salt industry. Several patents were granted, the first of which in Massachusetts, to monopolize salt. In other words, the process was a pure imitation, not even mixed with innovation, even if it was called a 'patent.'

The legal protection for artisans spread in the USA to encourage them to imitate and replicate what did not exist there or existed as commodities in the markets (colonial markets) but not internally produced in the colonies. Before independence, from the seventeenth century to the end of the eighteenth century, the American colonies were geographical areas rich in raw materials and suffering from severe shortages in the workforce and advanced manufacturing industries. Thus, the economic policy then was to export raw materials to Europe and import workforce (slaves predominantly) and finished industrial products.

This historical story shows us three things: first, how European culture was reflected in the American colonies and how its concepts of copycatting and imitation influenced them; second, how imitation and replication entered the USA from its inception as the exclusive meaning of innovation; and third, the nature of the unindustrialized economic situation in the American colonies gave birth to the political opinion of the leaders in the era of independence and encouraged imitation and copycatting with the aim of self-sufficiency. To reinforce the concepts referred to in the previous tale, we will cite another story related to the patent on steam-powered ships, which coincided with the US independence. It gives reasonable evidence of copying and imitation in the American culture before the independence of the State (in this chapter) and after the independence (in the next chapter).

The following lines provide an overview of the history of patents in the US colonies before the patent system was issued in 1790 and before the Declaration of Independence in 1776. We will re-examine patents and their systems in the US colonies that used to issue patent concessions to learn about the amount of copying, imitation, or originality in the patents granted in the colonies before the independence.

In 1606, King James I granted a patent to two companies to establish two colonies on the east coast of North America. The Virginia Company of London and the Virginia Company of Plymouth obtained a patent to settle in the new land overseas. This patent granted both companies the right to settle in specific areas and enact laws and regulations in the new colonies. Virginia was a corporate colony (private sector) before it became a royal colony of the British Crown in 1676 (6).

In 1619, individuals' first patents were granted to settle in the new colonies. William Fairfax was one of the first to receive a patent for a 200-acre land in Virginia, followed by a series of patents issued to individuals for different land areas ranging from 50 to more than 8,000 acres per capita. In 1625, 184 people were granted patents to settle in the new land (7).

In 1620, John Wyncop and John Peirce were granted a patent for the settlement and cultivation of Jamestown (the first British colony in the US) (8). In 1623, King James I granted a patent for aiding and assisting George Calvert to settle in Newfoundland (today in Canada), then called Avehn (6).

In 1628, John Endecott was granted a patent for land settlement from Weymouth to Abigail in New England (a group of colonies and states on the east coast of North America) (9).

The term 'patent' remained the common legal term for the royal privilege granted (to individuals or companies) to monopolize land, commodity, or commercial activity. The term was not related to innovation but remained in practice in England and its colonies until the term 'charter' replaced it in the colonies. All the old documents of the colonies refer to a patent, but it did not mean what we know today. In their book, *The Cambridge History of Law in America,* Michael Grossberg and Christopher Tomlins provided a full explanation of the terms used in the US colonies since the seventeenth century and the origin of the term 'patent' for land grants and 'charter,' which replaced it later to describe the granting of land after 1660. Nevertheless, the term 'patent' continued to be used after this date in the colonies. It became common in government records to describe the granting of land and the privilege of monopolizing the trade of goods (6).

The companies in the colonies were also established as per a 'patent' granted by the king. This was the case with the first companies in the colonies since 1555, such as the Muscory Company, the Merchant Adventurers Company, the Levant Company, and most notably the East Indian Company. Entire colonies were handed over to companies for settlement, profit from the colonies' activity, provide markets for them, and export the goods they produced (6). On this methodological foundation in granting patents, in 1641, Samuel Winslow was granted a ten-year patent by the Massachusetts Local Government Board to manufacture salt in state lands. Britain was then involved in an internal civil war, which reduced its authority over the colonies. Thus, the colonies took measures to grant patents on their own without referring to the British Crown. In 1641, John Jenny was granted a 21-year patent for making salt on 30 acres in Plymouth and selling each bushel of salt for 2 shillings. In 1648, John Winthrop was granted a patent for manufacturing salt in the colonies (10).

The study of the history of patents in the American colonies proves that patents were not based on any innovation or invention. They were essentially a privilege to monopolize a business or hold land. Even the colonies' salt industry (for example) was not based on innovations but on the transfer of the salt industry or commercial or industrial activity to the colonies, where industries were scarce (11). It was, then, a method of attracting and encouraging investment in well-known technologies, a process of stimulating, copycatting, and imitating the technologies available in Europe.

# References

1. C. K. Harley, in *New Perspectives on the Life and Work of Eric Williams* (University of Oxford, St. Catherine's College, Oxford University, UK, 2013).

2. L. Southwick, The London Cutler 'Benjamin Stone' and the Hounslow Sword and Blade Manufactories. *Arms & Armour* **6**, 12-61 (2009).

3. C. Welch, W. C. o. Cutlers, *History of the Cutlers' Company of London and of the Minor Cutlery Crafts: With Biographical Notices of Early London Cutlers. From 1500 to modern times.* (Printed privately for the Cutlers' Company, 1916).

4. W. A. a. L. Griswold, Donald W., "Saugus Iron Works: The Roland W. Robbins Excavations, 1948-1953, Edited by " (National Park Service. US Department of the Interior, Saugus, Massachusetts, USA, 1954).

5. T. Brown, *Popular Patents: America›s first inventions from the airplane to the zipper*. (Scarecrow Press, New Jersey, USA, 2000).

6. M. S. Bilder, in *The Cambridge History of Law in America: Volume 1: Early America (1580–1815),* C. Tomlins, M. Grossberg, Eds. (Cambridge University Press, Cambridge, 2008), vol. 1, pp. 63-103.

7. J. F. Fausz, College of William & Mary - Arts & Sciences, (1971).

8. P. M. Baker, "The Plymouth Colony Patent: Setting The Stage," (Pilgrim Society & Pilgrim Hall Museum, 2007).

9. M. Cheung, Rethinking the History of the Seventeenth-Century Colonial Ordinance: A. Reinterpretation of An Ancient Statute *Maine Law Review* **42**, (1990).

10. J. W. Cortada, *Rise of the Knowledge Worker*. (Butterworth-Heinemann, 1998).

11. K. W. Dobyns, *The Patent Office Pony: A History Of The Early Patent Office*. (Docent Press, Pennsylvania, USA, ed. 2nd, 2016).

# CHAPTER FOURTEEN

CHAPTER FOURTEEN

# AMERICAN IMITATION AND REPLICATION OF EUROPEAN INNOVATIONS SINCE THE DECLARATION OF INDEPENDENCE

George Washington (the first US President after independence, 1789-1797) and Thomas Jefferson (the first US Secretary of Finance after independence and the third US President, 1801-1809) sensed the danger of independence while being weak in terms of the industries coming from Europe. They wanted, by all means, to establish similar industries in the US colonies. Washington even tried to grant importation patent rights, but the vote was to support patents for industries within the colonies without importation. It was just an imitation of what was in Europe. In this way, the USA started its independence. Its political independence required independent economic production, which adopted the methodology and philosophy of imitation and replication to imitate products imported from Europe to the colonies.

Reflecting on the idea mentioned above, George Washington said on January 8, 1790, in his first annual address to the Congress after he was appointed President, "*I cannot forbear intimating to you the expediency of giving effectual encouragement as well to the introduction of new and useful inventions from abroad as to the exertions of skill and genius in producing them at home.*" (1)

Following declaring the independence of the US colonies on July 4, 1776, in Philadelphia, Pennsylvania, and before the patent system was issued in 1790, James Rumsey and John Fitch separately obtained several patents from different states (and some states issued a patent for both for the same invention). The invention was the manufacture of steamships that operate by a steam-powered engine fixed on board. Both inventors claimed the right of precedence in the invention to exclude the other from the patent monopoly. The political leaders of the independent US colonies, including President George Washington, intervened to resolve this dispute (2-4).

Back to the original rationale, this work was not an innovation. Both parties and the political leaders knew that. The process was to grant a franchise right to monopolize the manufacture of the copied or counterfeit product in the US colonies.

Around 80 years before the two conflicting parties claimed that innovation in the colonies, specifically in 1705, the French inventor Denis Papin placed a steam engine on a ship powered by oars mechanical propulsion. Using the steam engine instead of humans to move the oars, the model was first used in the Fulda River in France. Then the Englishman Thomas Newcomer experimented with the same idea but with a larger and heavier steam engine. These initial attempts by Babin and Newcomer led John Allen to obtain a patent in England in 1729, followed by Jonathan Hulls for the steamship with a change in the oars and the engine working mechanism. When William Henry visited Britain, he returned to the USA with the idea of manufacturing a steam-powered boat, but without applying for a patent on it. In 1783, the French Claude-François-Dorothée, marquis de Jouffroy d'Abbans, developed a model of an oarless steam-powered ship with a gear-like wheel connected to the steam engine that operated by water pumping when the gears connected to the engine rotate to move the ship. In Scotland, William Symington developed a model of a steam-powered ship (5-8).

This example shows that the patent in the USA was not related to innovation. The steamboat existed eighty years before granting the patent in the US colonies. It was a political process to encourage national industrialization and transfer foreign technologies to the new colonies. It was never an encouragement for invention and innovation.

To analyze the US patent system that coincided with the independence period, we refer to the Constitutional Convention, Article 1, Section 8, Clause 8, proclaimed in 1787. It referred to the acceleration of giving effective stimulus to introduce new and beneficial innovations coming from abroad. In other words, the US Constitution not only allowed but encouraged the piracy and transfer of foreign innovations to the new land. George Washington and some of the first legislators issued patents even on imports[8]. Although the Congress did not approve them, they were still permitted to be issued through the local house of each state, an organizational legislature entity junior to the Congress (2, 3).

In 1790, the US Patent Act was issued to establish a three-member committee to review the innovations filed. Thomas Jefferson was part of the team. Three patents were granted in the first year after the system was issued. The number

8 Importation patent: a legal protection document that grants an exclusive right to import certain goods for a specified period.

of patents reached 57 during the first three years. The term of legal protection for these first innovations was fixed at 14 years from the date of filing. However, the system changed in 1793, and the arbitration and examination of patents became not mandatory. The filing process comprised registering a patent without obligatory examination. In 1802, the USPTO set up records for these innovations and a headquarters for invention registration officers. The inventions had great importance in supporting the emerging US economy, then lacking advanced industries from Europe or the East. However, the USPTO was hit by a fire in 1836 and lost all its documents. In the same year, a new and revised patent system was issued, re-requiring examination as a criterion for granting any patent (2, 3, 9).

It is worth noting that the US laws drew upon the English laws and were infused with French revolutionary concepts. Before independence, all the US political leaders were English citizens in the colonies, and their relationship with the French was very close. Thomas Jefferson was the ambassador of the American colonies in France after independence, and he later became the third US President.

However, the US patent system was characterized by the low costs of filing and granting patents compared to European countries. These features encouraged artisans and workers to register their inventions, as indicated by Zorina Khan in her book published in 2006 about the democracy of innovations in the US in the eighteenth and nineteenth centuries. The USPTO filing fee was US$0.5 in 1790, and the patent issuance fees were US$2, plus 10 cents per page of the innovation specification (10). In 1990, Sokolov pointed out after analyzing the USPTO patent register during the 56 years from 1790-1846 that a remarkable proportion of these patents was for applying principles and mechanisms already known (11).

Although all innovation records were burned in 1836, some information about the first invention patented after the issuance of the system was available. Samuel Hopkins (1743-1818) was granted a patent three months after the system was announced for 14-year legal protection on the USX1 patent for producing potassium carbonate ($K_2CO_3$), known as pearl ash (due to its color), used in the baking industry and had other uses in the soap, glass, and ceramic industries. The innovation was about a method for a furnace to burn ash before boiling it in water that would help produce potassium carbonate (12, 13).

It was reported that that substance was used in Babylon for soap production since 2800 BC (about 5,000 years ago), more than four thousand years before Hopkins' so-called invention. Clay tablets found in Mesopotamia dating back to 2200 BC described the mechanism of its use in industries. It was used by the ancient Romans as well as found in Mexico a thousand years BC. It was produced in several ways, including passing water over types of stones to extract biocarbonite, and passing water and filtering it on some types of marine plants or ash (14). It was later imported to the US colonies from Europe (15). In 1998, Maxey pointed out that Hopkins was not an inventor genius, but a person of ordinary abilities. His invention was described as a recipe in a cookbook (12).

This product, 'pearl ash,' was present in the US markets before the invention as an import from Canada and widely used in the baking industry (15). The Canadian Angus MacDonnell filed a patent before Hopkins and applied for legal protection in Canada. After Hopkins reintroduced the invention in Canada, it was discussed. An expert indicated to the evaluation committee that the principle in the manufacturing process was the same for both inventions (12). So, the first US invention after independence was a historically replicated and contemporary counterfeit product in Canada. This fact shows us the profound impact of imitation and replication on the US economy since its inception.

It is also worth noting that the contents of many patents retrieved in the 1836 fire, which did not exceed 25% of the patents granted from 1790-1836 (as 75% of the patents were lost entirely), were incomplete. For example, the first ten patents retrieved contained little information, such as the inventor's name, grant date, and title (except for the first patent explained above) (16). Therefore, it is impossible to analyze the extent of innovation or imitation therein.

# References


1. G. Washington, *State of the Union: Selected Annual Presidential Addresses to Congress, from George Washington, Abraham Lincoln, Franklin Roosevelt, Ronald Reagan, George Bush, Barack Obama, and Others.* (Red and Black Publishers, St Petersburg, Florida, USA, 2009).
2. F. Neumeyer, Contribution to the history of modern patent legislation in the United States and in France. *Scandinavian Economic History Review* **4**, 126-150 (1956).
3. C. A. Nard, A. P. Morriss, Constitutionalizing Patents: From Venice to Philadelphia. *Review of Law & Economics* **2**, 223-321 (2006).
4. C. W. Hackensmith, JOHN FITCH: A PIONEER IN THE DEVELOPMENT OF THE STEAMBOAT. *The Register of the Kentucky Historical Society* **65**, 187-211 (1967).
5. H. W. Dickinson, *A Short History of the Steam Engine (Cambridge Library Collection - Technology).* (Cambridge University Press, Cambridge, England, 2011).
6. T. Crump, *A Brief History of the Age of Steam: From the First Engine to the Boats and Railways.* (Running Press, Philadelphia, Pennsylvania, USA, 2007).
7. F. Klemm, *A History of Western Technology.* (Iowa State University Press, 1991).
8. B. Clark, *Steamboat Evolution.* (Lulu.com, 2007).
9. K. W. Dobyns, *The Patent Office Pony: A History Of The Early Patent Office.* (Docent Press, Pennsylvania, USA, ed. 2nd, 2016).
10. B. Zorina Khan, *The Democratization of Invention Patents and Copyrights in American Economic Development, 1790–1920.* (Bowdoin College, Maine, 2009).

11. K. L. Sokoloff, B. Z. Khan, The Democratization of Invention During Early Industrialization: Evidence from the United States, 1790-1846. *The Journal of Economic History* **50**, 363-378 (1990).

12. S. Hopkins, D. W. Maxey, Samuel Hopkins, the Holder of the First U.S. Patent: A Study of Failure. *The Pennsylvania Magazine of History and Biography* **122**, 3-37 (1998).

13. M. Nascimento, The First Patents and the Rise of Glass Technology. *Recent Innovations in Chemical Engineering (Formerly Recent Patents on Chemical Engineering)* **9**, 20-30 (2016).

14. G. Létourneau, J. Sames, Ash to Cash – The Untold Story: Nature's Burnt Offering to 19th Century Settlers. *Histoire Québec* **18**, 25-30 (2013).

15. C. A. Browne, Historical notes upon the domestic potash industry in early colonial and later times. *Journal of Chemical Education* **3**, 749 (1926).

16. H. H. d. S. Nascimento, C. N. d. Santos, C. H. A. Alves, M. L. F. Nascimento, The X-Patents. *World Patent Information* **53**, 1-13 (2018).

# CHAPTER FIFTEEN

CHAPTER FIFTEEN

# MODELS OF THE US IMITATION IN WORLD WAR II AND THE COLD WAR

This chapter includes selected notable innovations or scientific breakthroughs that resonated significantly in the US and abroad as proclaimed US inventions. We will analyze the originality of the innovation or the scientific idea on which the innovation relied to know the amount of copying and imitation in the roots of major US innovations and scientific breakthroughs.

With regard to the invention of the lighting lamp, Thomas Edison (1847-1931) filed a patent in the USA in 1878 under the title "Electric Light" and obtained patent No. 223898 in 1880 (1). Ernest Freeberg, author of *The Age of Edison: Electric Light and the Invention of Modern America*, indicated that Edison was not the inventor of the electric light. Edison borrowed (stole as per the author's description) the idea from other researchers who were working on electric light, including the English scientist Joseph Swan (1828-1914), who preceded Edison by years in developing the concepts of electric light production and proved this by reviewing previous studies before the British court when he sued Edison for infringing on his innovation (2). However, Edison manufactured the idea and established a company (Edison Electric Light Company) in the USA to produce electricity in 1878, before Swan established his company (Swan United Electric Light Company) in 1881. Then Edison offered Swan to join him and merge the two companies to resolve the dispute, and Swan agreed and established the Edison & Swan United Electric Light Company. Then the company entered into a series of mergers and partnerships until it became the General Electric (GE) Company and was introduced in an initial public offering (IPO) in the US Dow Jones market for industry (it still exists today with annual revenues of nearly US$ 100 billion) (3-5). Frimberg also pointed out that most US innovations adopted this approach, i.e., to take others' scientific innovations and manufacture devices based on them and then re-market them to the world (2).

Several sources indicate that electric light was invented in 1816 (some references refer to 1817) by Humphry Davy of the Royal Institution of Great Britain (4, 6).

Concerning the invention of cars that run on fossil fuels, they were not a US innovation, as Barack Obama, the forty-fourth US President (2009-2017), claimed in one of his speeches on February 24, 2009[9]. It was not an innovation by Henry Ford (1863-1947), the founder of the Ford Motor Company in 1903, and the Ford T-Model producer in 1908 (7, 8). The fossil-fuel-powered car was an innovation of Karl Friedrich Benzes (1844-1929), granted patent number 37435 in 1885 in Germany. He produced its prototype in 1886 and later developed the Mercedes-Benz that exists today (9-11).

Concerning the invention of telephone, Alexander Graham Bell (1847-1922), a Canadian of British origin, invented it when he obtained patent No. 174465 in 1876 from the USPTO entitled "apparatus for transmitting vocal or other sounds telegraphically" (12). The Italian inventor Antonio Meucci (1808-1889) from Florence, Italy, invented the telephone in 1860 and received patent No. 36419 from the USPTO in 1862 (13, 14). Also, Bell invented the telephone while in Canada. Thus, it is not fair to say that the telephone is a US invention despite registering it with the USPTO (15).

With regard to the invention of the aircraft, the brothers Orville (1871-1948) and Wilbur Wright (1867-1912) filed a

9 https://www.bloomberg.com/news/articles/2009-03-05/the-car-a-u-dot-s-dot-invention-nein

patent in 1903 entitled "Flying Machine" and they were granted a patent No. US821393 in 1906 from the USPTO (16, 17). However, the plane was invented before that. Shivkar Bapuji Talpade (1864-1916) developed a model plane and flew it in 1895 in Bombay, India. The French of Brazilian descent Alberto Santos-Sumont (1873-1932) flew on a plane before the Wright brothers in France. His model was attached to a wheel (18). The German immigrant to the US Gustave Whitehead (1874-1927) developed a model for flying an airplane in 1901 (19). The Wright brothers may have been the first to patent an aircraft in the USPTO, but they were not the first to invent the aircraft. More than a thousand years before the Wright brothers' patent, the Arabic scientist Abbas ibn Firnas (810-887 CE) documented the construction of an aircraft flying in Cordoba, Andalusia, in the ninth century CE (20).

The US space program is the most significant imitation project in size and capabilities in the contemporary world. The USA made a determined and organized attempt to allocate considerable funds to it to simulate, copy, and imitate Soviet innovations in the world of space and missiles, starting from imitating the Soviet R-7 intercontinental ballistic missiles (ICBM) and placing an animal and then a man on a spacecraft.

The USA spent more than US$400 billion on the space project only from 1958 to 1962 under President Dwight Eisenhower (1953-1961) and John Kennedy (1961-1963 (21). This period was a turning point in the history of the US copying, imitation, and simulation of Soviet innovations, after the Soviet success in sending the first spacecraft, Sputnik 1, in October 1957. At the end of 1962, government spending on the US space program doubled to become more than 1% of total government spending. Under President Lyndon Johnson (1963-1969), the government spending on the space program reached 4.4% of total government spending in 1966 (21). Allocating a high percentage of the government budget to a project that did not directly relate to any of the people's concerns, such as health, education, and housing, was a great sacrifice and required a lot of advocacy before the people and the Senate to justify.

However, the shock received by the US after the launch of a satellite in 1957 (rotates around the Earth and sends radio pulses) by the Soviet Union in the Sputnik 1 project, was a tragedy for the US leadership and people. The biggest tragedy was sending an animal (a dog named Laika) in the Sputnik 2 project (22). These shocks led to the establishment of the US National Aeronautics and Space Administration (NASA) less than three months later in 1958, and the issuance of the National Aeronautics and Space Act (23). It also led to a change in the education system with the issuance of the National Defense Education

Act (24). This date marked the start of the US largest project of copycatting and imitation after the Industrial Revolution. Imitation and copycatting for the US leadership was the strategic solution for obtaining the Soviet technical advantage. Indeed, after harnessing all these possibilities of imitation and replication, the USA succeeded and maintained its international weight and influence. What if the USA decided not to imitate the Soviet space project from the beginning? We can only imagine the different course of history! Perhaps the USA would have become part of the Soviet Union! In this book, we do not limit the concept of imitation and copycatting to negativity. There may be positive aspects to it and an easier solution for those who missed innovation.

The gap between the Soviet superiority in space and missile innovations was huge compared to the US technology (25, 26). The range of the first ICBM the Soviets launched in 1957 (the R 7 missile) was 8,800 km, while that of the most potent US missile at that time, the Atlas A, was only 1,000 km. We should note that the radius of the Earth is less than 8,000 km, i.e., the Russian missile could have hit anywhere on the surface of the Earth. In contrast, the US missile could not cross the Pacific or the Atlantic Ocean or hit Latin American targets from a launchpad in the USA territories. The technological gap was too big in favor of the Soviets.

In space, in addition to sending the first satellite by the Soviets (Sputnik 1) and sending the first living creature to space in Sputnik 2 in 1957, the Soviets reached the moon in 1959 on the Luna 2 spacecraft and sent the first human, Yuri Gagarin, to space in 1961 aboard the Vostok 1. The Russians brought the first sample from space to Earth aboard the Luna 16 spacecraft, sent the first spacecraft, Venera 4, to Venus in 1967, and the Mars 2 spacecraft to Mars in 1971 (22, 27-29). Keeping up with and simulating the innovations of the Soviets was a dream for the US leaders. It took the USA more than 13 years of continuous copying, imitation, and simulation efforts to advance in some areas of space at the beginning of the seventies of the twentieth century. To showcase the imitations of the US innovations in the space field, the Explorer 1 satellite, launched in 1958, was an imitation of the Russian Sputnik 1 satellite. The entire US Mercury Program was an attempt to imitate a series of successful Soviet innovations, including the spacecraft that carried the chimpanzee Ham, which was launched in 1961; it was an imitation of Sputnik 2, which carried a dog. The Mercury Red Stone spacecraft launched in 1961 as the first US spacecraft to carry the second human, Alan Shepard, to space was an imitation of the success of the Soviet spacecraft Vostek 1 (30-35).

These are only a tip of the iceberg. This chapter was not designed to list all the products of the US imitation, but to provide an overview of the extent to which the USA adopted and relied on imitation and replication as a strategy, policy, and a successful method to keep pace with Europe in the eighteenth century after the Declaration of Independence (as we explained at the beginning of this Part). This approach remained a successful tactic in the Cold War with the Soviets. This was preceded by attempts to imitate German innovations in radars and submarines in World War II. It is worth noting that the USA received a shock no less than the tragedy of Sputnik in World War II in front of the German Nazis, which made it harness all its capabilities and establish a secret laboratory at Harvard University with 800 scientists from all over the country to identify the secret of German technical superiority in radar science after the Germans destroyed all the US Air Fleet sent to back its allies in Europe. The Americans were surprised that the German radar technology recognized the planes 200 km before approaching the coasts of Europe, and thus bombed them with such high accuracy. The German planes at that time had radars that enabled them to detect enemy planes at a distance of 5km in the air, which facilitated the destruction of the US Air Fleet (36).

As a clear example of the Allies' desire to imitate and copy in World War II, Britain and the USA carried out a raid on the shores of Europe to steal, not destroy, a German radar. They wanted to obtain a sample of the superior German radars to reproduce and replicate it.

Copycatting and imitation were thus an effective strategy even in battles and wars: the opponent could be prevented from gaining military supremacy by re-possessing the same superior weapon the opponent had, an old strategy in war. While innovation was a solution to overcoming the military superiority of the opponent, imitation and copycatting had always been more accessible and practical solutions than an innovation that had a high risk of failure. Imitation is a replication of successful innovation, tried and proven beneficial. As for innovation, and even if it succeeds in manufacturing or production, it may not succeed in the practical use on the battlefield, let alone the high costs of innovation compared to copying and imitating existing innovation.

Going back to the previous example, James Baxter, the head of the US Strategic Services, described the black metal box that the British transferred to the Americans after the radar theft Tizard Mission as "*the most valuable cargo ever brought to our shores.*"The mission or German radar theft raid was named after its leader Henry Tizard, chairman of the British Aeronautical Research Committee (37).

The advanced German radar, "Freya radar," had a range of 200 km. One thousand units were produced in 1939. It was superior to all existing technologies at that time. When these radars were installed on the coastal cities of Western Europe after the Germans occupied France, the Germans could detect the movement of British royal planes while at their bases in Britain and monitor the US planes hundreds of kilometers before approaching the shores of Western Europe (38). The radar name was derived from an old Scandinavian and German goddess (old myths) called Freya, the goddess of war, beauty, and fertility. They believed that half of those who die in war were received by Freya, and the other half went to another god called Odin, the god of possessions, who received them in a hall of martyrs called Valhalla. However, we do not know the connection between the goddess of fertility in mythology and German radar technology (39)!

Our review of space technologies does not aim to glorify the Soviets at the expense of the Americans, nor vice versa. It is just a re-reading and an analysis of the historical facts to benefit from the experiences of the extinct and existing nations. The origin of the idea of the space conquest was authors of idealistic literary works (utopias) not based on science or technology or accompanied by executive programs applicable on the ground (40, 41). Nikolai Fedorov (1829-1903) and Konstantin Tsialkovsky (1857-1935) referred in their literary works to fantasies about human movement through space and the manufacture of missiles to travel to other planets (42, 43). In the twenties of the last century, literary works about space, stars, and the human presence thereon became a part of the novel global culture. Then, in 1937, literary science fiction works were banned in the Soviet Union during the era of Joseph Stalin (1922-1953) (40).

However, during the Nazi era, the Germans developed the first operational program to invade space with the A-4 missile, later developed into a version called Vengeance V-2, which weighed 13 tons and carried a one-ton warhead. The task was entrusted to the Peenemünde Army Research Center. The Germans produced more than 3,000 V2-class missiles since 1942, with a range of 320 km. The missile could cross the mesosphere and enter the thermosphere, whose height starts from 85 km from the Earth's surface to the Kármán line, 100 km from the Earth's surface. Although the V-2 could not enter outer space and exceed the thermosphere, which is 690 km from the Earth's surface, it was the first rocket to reach this distance and approach the exosphere (44-48).

To do right, thanks are due to the efforts made by German scientists during the Nazi era for the access to space by the Russians and Americans in the fifties, sixties, and early seventies. The process of re-manufacturing the same product and then developing it brought the Soviets and then the Americans to space and ICBMs. Strangely, the defeated in war taught the victor how to succeed. How could the imitator outperform the innovator? The later chapters will address this.

To clarify the background of innovation in Nazi Germany and how it was transmitted (or copied) afterward to Russia and then the US, we refer to the Peenemünde Army Research Center, where a team of German scientists led by Wernher Van Braun (1912-1977) and Herman Julius Oberth (1894-1989) manufactured the A-4 missile followed by its upgraded version V-2 as we mentioned above. These missiles destroyed many sites in Britain (48, 49).

After the war ended, the Soviets, as well as the Americans, seized the stock of knowledge and the German scientists who had survived the Second World War. Braun moved with a group of scientists to the USA. The Soviets, working more seriously, managed to complete some of the Nazi German projects related to the V-2 space program. Soviet President Stalin asked the scientist Sergei Korolev (1907-1966) to replicate the German A-4 missile. Korolev offered to develop the German version, but Stalin refused and insisted on *reproducing* the same version, and keeping the development later. This demand marked the beginning of Soviet supremacy in space science, as it took Korolev two years to reproduce the A-4 rocket. This operation led the

Soviets, led by Korolev, to the production of the R-7 ICBM and the Sputnik 1 spacecraft in 1957. The scientific foundations used by the Germans were the same used later by the Soviets to reach space (41).

As per the exact copycatting and imitation methodology, Britain and Canada, supported by NASA, launched their first satellites in 1962. The US satellite Alouette 1 was assembled in Ottawa, Canada, and then launched from the military airbase in Vandenberg, California (50, 51). The British launched the Ariel 1 satellite (manufactured in the US) from a launchpad made by the US on British soil. The Italians sent the San Marco 1 satellite into space in 1964 (52). In 1965, France manufactured and launched the Asterix satellite (manufactured in France) from the Diamant Rocket launchpad in Algeria. However, the French made the satellite and launchpad by simulating the US model (53).

Despite the different external shapes of satellites and launchpads used in these countries, the whole process was pure copycatting and imitating the original idea of the Sputnik 1. The French also sent Félicette, the first cat, to space in a simulation operation of Sputnik 2. As for the rocket used to launch satellites, these countries introduced only minor improvements to the Nazi A-4 and V-2 missiles. The methodology of copying and imitation was the means that brought the communist and capitalist countries into space. It seems that philosophy and economical and social intellectual methodology had nothing to do with the technical progress of both intellectual currents, i.e., communism and capitalism. The author may claim that the methodology of copycatting, imitation, and simulation lies behind technical excellence and knowledge progress. The economic philosophies that the world has been grappling over for more than a century have nothing to do with that.

# References


1. T. A. Edison, in *US Patent Office,* U. P. Office, Ed. (USA, 1880).
2. E. Freeberg, *The Age of Edison: Electric Light and the Invention of Modern America.* (Penguin Publishing Group, 2013).
3. G. Wise, Innovation: Swan's way: A study in style: With the birth of the electric light bulb, two modern styles of inventing emerged: Thomas Edison's and Joseph Swan's. *IEEE Spectrum* **19**, 66-71 (1982).
4. S. Ruston, Humphry Davy in 1816: Letters & the lamp. *Wordsworth Circle* **48**, 6-16 (2017).
5. D. E. Nye, Edison and the shadow side of artificial light. *Nature*, 326-327 (2019).
6. Anonymous, *The Story of Sir Humphry Davy and the Invention of the Safety-Lamp. (Lessons From Noble Lives).* (Sagwan Press, 2018).
7. R. Banham, *The Ford Century: Ford Motor Company and the Innovations that Shaped the World.* (Artisan, 2002).
8. R. M. Langworth, *The Complete History of Ford Motor Company.* (Beekman House, 1987).
9. S. J. C. Nixon, *The Invention of the Automobile - (Karl Benz and Gottlieb Daimler).* (Edizioni Savine, 2016).
10. K. Benz, in *German Patent Office,* G. P. Office, Ed. (Co Rheinische Automobil und Motorenfabrik AG, Germany 1886).
11. C. Dowlen, Automobile design history: What can we learn from the behavior at the edges? *International Journal of Design Creativity and Innovation* **1**, (2013).
12. A. G. Bell, in *US Patent Office,* U. P. Office, Ed. (USA, 1876).
13. A. Campanella, Antonio Meucci, The Speaking Telegraph, and The First Telephone. *Acoustics Today* **3**, (2007).

14. A. Meucci, in *US Patent Office,* U. P. Office, Ed. (USA, 1862).

15. E. S. Grosvenor, M. Wesson, R. V. Bruce, *Alexander Graham Bell: The Life and Times of the Man who Invented the Telephone.* (Harry Abrams, 1997).

16. O. W. Wright, Wilbur in *US Patent Office,* U. P. Office, Ed. (USA, 1906).

17. O. Wright, *How We Invented the Airplane: An Illustrated History (Dover Transportation).* (Dover Publications, Mineola, New York, USA, 1988).

18. A. Nahum, Getting it off the ground. *Nature* **426**, 765-766 (2003).

19. C. KLEIN. (History, 2018).

20. E. A. Jamsari *et al.*, Ibn Firnas and His Contribution to the Aviation Technology of the World. *Advances in Natural and Applied Sciences* **7**, 74-78 (2013).

21. (US Office of Managment and Budget).

22. A. P. Cracknell, C. A. Varotsos, Editorial and cover: Fifty years after the first artificial satellite: from Sputnik 1 to ENVISAT. *International Journal of Remote Sensing* **28**, 2071-2072 (2007).

23. U. S. o. America, Ed. (1958).

24. U. S. o. America, Ed. (1958).

25. B. G. Shreve, THE US, THE USSR, AND SPACE EXPLORATION, 1957-1963. *International Journal on World Peace* **20**, 67-83 (2003).

26. K. Werth, A Surrogate for War—The U.S. Space Program in the 1960s. *Amerikastudien / American Studies* **49**, 563-587 (2004).

27. S. Gerovitch, *Voices of the Soviet Space Program: Cosmonauts, Soldiers, and Engineers Who Took the USSR into Space (Palgrave Studies in the History of Science and Technology).* (Palgrave Macmillan, London, England, 2014).

28. C. River, *Soviet Russia's Space Program During the Space Race: The History and Legacy of the Competition that Pushed America to the Moon.* (CreateSpace Independent Publishing Platform, 2015).

29. A. A. Siddiqi, *Sputnik and the Soviet Space Challenge.* (University Press of Florida, Gainesville, FL, USA, 2003).

30. J. Bell, *The Space Book: From the Beginning to the End of Time, 250 Milestones in the History of Space & Astronomy.* (Sterling New York, USA, 2013).

31. D. Baker, *The History of Manned Space Flight.* (Outlet, 1986).

32. T. SPITZMILLER, *THE HISTORY OF HUMAN SPACE FLIGHT.* (University Press of Florida, Gainesville, 2017).

33. M. J. Neufeld, *Spaceflight: A Concise History.* (The MIT Press, Cambridge, MA, USA, 2018), vol. MIT Press Essential Knowledge series.

34. R. D. Launius, *The History of Space Exploration Discoveries from the Ancient World to the Extraterrestrial Future.* (Thames & Hudson, London, England, 2018).

35. G. Kranz, *Failure Is Not an Option: Mission Control From Mercury to Apollo 13 and Beyond.* (Simon & Schuster, New York, USA, 2009).

36. V. Bush, *Science the endless frontier.* (United States Government Printing Office, 1945).

37. D. Zimmerman, H. T. Tizard, *Top Secret Exchange: The Tizard Mission and the Scientific War.* (Alan Sutton Pub., 1996).

38. F. P. V. Miller, Agnes F.; McBrewster, John, *Freya Radar: Early Warning Radar, World War II.* (Alphascipt Publishing, 2010).

39. A. Orchard, *Dictionary of Norse Myth and Legend.* (Cassell, 1997).

40. I. Kohonen, The Space Race and Soviet Utopian Thinking. *The Sociological Review* **57**, 114-131 (2009).

41. J. B. West, Historical aspects of the early Soviet/Russian manned space program. *Journal of Applied Physiology* **91**, 1501-1511 (2001).

42. G. M. Young, *The Russian Cosmists: The Esoteric Futurism of Nikolai Fedorov and His Followers.* (Oxford University Press, Oxford, England, 2012).

43. K. E. Tsiolkovsky, *Selected Works of Konstantin E. Tsiolkovsky.* (University Press of the Pacific, Honolulu, Hawaii, USA, 2004).

44. W. Dornberger, *V-2 The Nazi Rocket Weapon.* (Ballantine, New York, USA, 1954).

45. G. P. Kennedy, *Germany's V-2 Rocket.* (Schiffer Publishing, Atglen, PA, USA, 2006).

46. M. J. Neufeld, *The Rocket and the Reich: Peenemunde and the Coming of the Ballistic Missile Era.* (Free Press, New York, USA, 1994).

47. N. Longmate, *Hitler's Rockets: The Story of the V-2s.* (Skyhorse, Onderstepoort, Pretoria, South Africa, 2009).

48. M. B. Petersen, *Missiles for the Fatherland: Peenemünde, National Socialism, and the V-2 Missile (Cambridge Centennial of Flight).* (Cambridge University Press, Cambridge, UK, 2011).

49. M. J. Neufeld, *Von Braun: Dreamer of Space, Engineer of War* (Vintage, Visalia, CA, USA, 2008).

50. E. Howell, *Canadarm And Collaboration: How Canada's Astronauts And Space Robots Explore New Worlds.* (ECW Press, Toronto, Ontario, Canada, 2020).

51. D. H. Jelly, *Canada 25 years in space.* (Polyscience Publications Laval, Québec, Canada, 1988).

52. H. Massie, M. O. Robins, *History of British Space Science.* (Cambridge University Press, Cambridge, 1986).

53. J. a. R. Krige, A., "A History of the European Space Agency: 1958 – 1987,» (EESA, 2000).

# Word Index

# Word Index